Yanomami
A forest people

Yanomami
A forest people

William Milliken AND Bruce Albert
WITH Gale Goodwin Gomez

ILLUSTRATIONS BY
Jane Rutherford

The Royal Botanic Gardens, Kew 1999

First published 1999

ISBN 1 900347 73 3

The authors:

William Milliken, ethnobotanist, associate research fellow at the Royal Botanic Garden, Edinburgh, EH3 5LR, UK. Has worked with the Yanomami since 1993.
email: W.Milliken@mcmail.com

Bruce Albert, anthropologist, senior researcher at IRD (Institut de Recherche pour le Développement), 209–213 rue La Fayette, 75480 - Paris Cedex 10, France. Has worked with the Yanomami since 1975.
email: bruce.albert@ibm.net

Gale Goodwin Gomez, ethnolinguist, lecturer in the Department of Anthropology and Geography, Rhode Island College, Providence, RI, 02908-1991 USA. Has worked with the Yanomami since 1984.
email: galemail@aol.com

The illustrator:

Jane Rutherford, artist and teacher, living and working in Scotland. Has painted among the landscapes and people of northern Amazonia in 1991 and 1994.

Cover design by Jeff Eden, page make-up by Media Resources, Information Services Department, Royal Botanic Gardens, Kew

Production Editor: S. Dickerson

Printed by in Great Britain
by B·A·S Printers Limited
Stockbridge, Hampshire

CONTENTS

ABSTRACT

Many aspects of the social organization, rituals and material culture of the Yanomami Indians of Brazil and Venezuela have been described and discussed in great detail. Much, however, still remains to be discovered about their interactions with the plant world. In this book a general overview of their use and knowledge of plants is presented, based on data collected from the Brazilian Yanomami (Demini, Balawaú and Xitei regions), who live in the northern part of the State of Amazonas and western part of the State of Roraima. The information is discussed in the context of the existing literature on Yanomami ethnobotany, and presented in the context of Amazonian ethnobotany in general. It is shown that a very substantial, and possibly exceptional, diversity of species is used by the Yanomami for a broad range of applications, but that in most cases the species employed and the uses to which they are put correspond closely to those which have been recorded amongst other Amazonian peoples.

RESUMO

Embora já existam descrições e discussões detalhadas sobre muitos aspectos da organização social, dos rituais e da cultura material Yanomami do Brasil e da Venezuela, muito tem ainda que ser descoberto a respeito de sua interação com o mundo vegetal. Neste livro apresentamos uma visão geral do uso e conhecimento florístico dos Yanomami, com base em dados coletados dos Yanomami do Brasil (regiões de Demini, Balawaú e Xitei), no norte do estado do Amazonas no oeste leste do estado de Roraima. Discutimos essas informações tanto no contexto da literatura etnobotânica Yanomami quanto no panorama geral da etnobotânica amazônica. Elas demonstram que os Yanomami se utilizam de uma grande diversidade de espécies, possivelmente excepcional, num amplo leque de aplicacões. Entretanto, na maioria dos casos, as espécies empregadas e seus usos correspondem de perto aos que foram registrados entre outros povos da Amazônia.

RÉSUMÉ

Bien que les aspects les plus divers de l'organisation sociale, des rituels et de la culture matérielle des Yanomami du Brésil et du Venezuela aient fait l'objet de nombreuses descriptions et discussions, beaucoup reste à découvrir de leur interaction avec le domaine végétal. Cet livre présente un aperçu général des connaissances botaniques yanomami basé sur des données collectées chez les Yanomami du Brésil (régions de Demini, Balawaú et Xitei), qui vivent dans le nord de l'état d'Amazonas et l'oest de l'état de Roraima. Ces données sont à la fois discutées dans le contexte de la littérature l'ethnobotanique yanomami et dans celui de l'ethnobotanique amazonienne en général. Elles démontrent que ces Indiens utilisent une considérable, voire une exceptionnelle diversité d'espèces végétales pour une large gamme d'applications. Toutefois, dans la plupart des cas, les espèces utilisées et leur usage correspondent étroitement à ceux qui ont été enregistrés dans d'autres groupes amazoniens.

INTRODUCTION

The Yanomami

The Yanomami are one of the few indigenous groups of the Amazon whose name is well known, both within the countries in which they live and abroad. For the last few decades they have been known principally for their controversial and unjustified reputation as a "fierce people" (Chagnon, 1968), and then more recently as the victims of the devastating and widely publicized consequences of the gold-rush which swept through their lands (in Brazil) at the end of the 1980s and the beginning of the 1990s. In spite of the fact that the Yanomami have been the subject of innumerable studies, anthropological and otherwise, these widely publicized issues have tended to draw attention away from the aspects of their society and way of living for which other indigenous peoples, in less dramatic or tragic circumstances, have been celebrated. One such aspect is their remarkable knowledge of the forest in which they live, of the species which inhabit it, and of the uses to which they may be put.

The Yanomami are a hunter-gatherer and swidden horticultural society of the tropical forests of the west of the Guiana (Guayana) Shield. They occupy a territory of approximately 192,000 km², spanning both sides of the Brazil-Venezuela frontier, and constitute a cultural and linguistic group composed of four or five sub-groups of the same linguistic family (see Migliazza, 1972; Ramirez, 1994). The overall Yanomami[1] population is around 26,000 people. The western Yanomami make up the greater part of this population (56%), followed by the eastern Yanomami (25%), the Sanima (14%) and the Ninam/Yanam (5%).

Yanomami villages generally consist of a large communal round house or, in the case of the Sanima and sometimes the Ninam/Yanam, a group of smaller rectangular houses. Each community considers itself economically and politically autonomous, and its members generally prefer to marry among themselves. All, however, maintain relations of matrimonial, ceremonial and economic exchange with neighbouring groups, who are considered political allies against other multi-community units of the same type. These units partially superimpose themselves to form a socio-political network which links all of the Yanomami villages from one extremity of their territory to the other.

Since the Yanomami are genetically, anthropometrically and linguistically separate from their neighbours such as the Ye'kuana (Caribs), geneticists and linguists have concluded that they are the descendants of an indigenous group which has remained in a state of relative isolation for a very long time. Having already established themselves as a distinct linguistic group ('Proto-

---

[1] Yanoama, Yanomamö and Yanomama have also been used in the literature as collective terms for these groups.

1

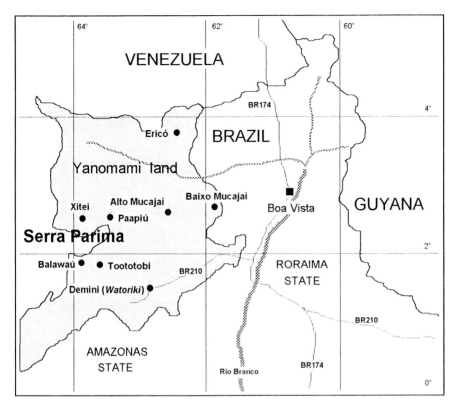

Figure 1. Location of the study sites. The stippled area represents Yanomami territory in Brazil.

Yanomami') approximately 2,500 years ago, the Yanomami would have occupied the Orinoco-Parima interfluve about 1,000 years ago, and there initiated the process of internal differentiation about 700 years ago, which ended with the languages and dialects of today (see Holmes, 1995; Migliazza, 1982; Neel *et al.*, 1972; Spielman *et al.*, 1979).

According to Yanomami oral tradition and to the earliest historical documents which mention this group, the centre of their territory is situated in the Serra Parima, which divides the waters of the upper Orinoco (Venezuela) and the upper Parima (Roraima, Brazil). This is still the most populated part of the Yanomami area. The dispersal of the population from that region in the direction of the surrounding lowlands probably began during the first half of the 19th century, after the colonial penetration of the upper Orinoco, Branco and Negro rivers in the latter half of the 18th century. The current configuration of the Yanomami territory has its roots in these past migratory movements.

This geographical expansion was made possible from the 19th century on to the beginning of the 20th century by strong demographic growth (Chagnon, 1974; Hames, 1983a; Kunstadter, 1979; Lizot, 1988). Various anthropologists consider this growth to have been caused by the economic transformations induced by the acquisition of new cultivars and metal tools through exchanges and wars with neighbouring indigenous groups (Caribs to the north and east; Arawaks to the south and west), who themselves maintained direct contact with the white frontier. The progressive emptying of the territories of these groups, which were devastated by their contact with regional society during the 19th century, also provided favourable conditions for the process of Yanomami expansion (Albert, 1985, 1990; Chagnon, 1966; Colchester, 1984; Good, 1995; Hames, 1983a; Smole, 1976).

The Yanomami in Brazil

The Yanomami population in Brazil was recently estimated as approximately 11,000 (DSY/RR - FNS, 1999). This population occupies the region of the upper Rio Branco (western Roraima) and the left side of the Rio Negro (northern Amazonas). The eastern Yanomami (from which most of the original field data in this book came) predominate in Brazil, with more than 5000 people. The history of contact between this indigenous society and the national society presents complex and regionally heterogeneous aspects, due to the successive economic frontiers which have penetrated their territory since the early 20th century, and with which they continue to coexist in a diversity of local combinations.

The Yanomami in Brazil first experienced direct contact with representatives of the national society (hunters, *balata* latex and *piaçaba* fibre collectors, soldiers working on the Boundary Commission CBDL, agents of the Indian Protection Service SPI, etc.) or with foreign travellers between the 1910s and the 1940s (see Albert, 1985). Between the 1940s and the mid-1960s, the opening of some SPI posts and, especially, of various evangelical and Catholic missions, established the first points of permanent contact in their territory. These posts constituted a network of foci for sedentarization, being regular sources of supply of manufactured goods (and also of lethal epidemics).

In the 1970s and 1980s the national development projects of the Brazilian State began to submit the Yanomami to increasingly intense forms of contact with the expanding regional economic frontier, principally in the west of the Roraima region: roads, colonization projects, ranches, sawmills, military bases and the first informal mineral prospecting sites (*garimpos*). These contacts provoked an epidemiological shock on a massive scale, causing heavy demographic losses, general sanitary degradation and serious social destructuring.

The two principal forms of contact initially experienced by the Yanomami – firstly with the extractivist frontier and then with the missionary frontier – coexisted until the beginning of the 1970s as the dominant external influence in their territory. However, the 1970s were marked (particularly in Roraima) by the implantation of development

Figure 2. *Garimpeiros* in the Yanomami territory (on the Rio Uraricoera) in 1989, at the height of the gold-rush.

Figure 3. The abandoned BR210 highway, close to the village of *Watoriki*.

projects under the auspices of the national integration plan (Plano de Integração Nacional) launched by the incumbent military governments. These took the form of the opening of a section of the BR210 Perimetral Norte highway (1973–76), and of public colonization programmes (1978–79) which invaded the southeast of the Yanomami lands. At the same time, a project surveying the resources of the Amazon (RADAM) in 1975 reported the existence of important mineral reserves in the region (see Ramos and Taylor, 1979). The publicity given to the potential wealth of the Yanomami area stimulated a progressive invasion of mineral prospectors at the end of the 1980s, escalating into a full-scale gold-rush in 1987. Over 80 clandestine airstrips were opened on the upper reaches of the main tributaries of the Rio Branco between 1987 and 1990, and the estimated number of prospectors in the area rose to 30–40,000 – five times the indigenous population in Roraima State (see MacMillan, 1995). Although the intensity of this gold-rush diminished greatly at the beginning of the 1990s, nuclei of prospecting still exist in the Yanomami area, and continue to act as sources of violence and severe sanitary and social problems.

The expansion of the mineral prospecting 'frontier' has tended, since the 1980s, to supplant the previous forms of contact which the Yanomami experienced with surrounding society, to the extent of relegating the development projects of the 1970s to a level of secondary importance. This does not imply, however, that other economic activities (agriculture, logging and ranching enterprises, industrial mining etc.), whether incipient or already fully active, may not constitute a new threat to the integrity of Yanomami lands in eastern Roraima in the near future, in spite of their limits having been legally recognized by Presidential decree in November 1991 and fully registered in May 1992.

Thus, as well as the persistent small scale-illegal prospecting interests in the region, it should be noted that the Yanomami territory is almost completely covered by official prospecting rights or bids which have been registered with the National Mineral Production Department (DNPM) by public, private, national and multinational mining companies. Furthermore, the colonization projects launched in 1978–79 in the extreme southeast of the Yanomami territory have created a population front which is tending to expand into the indigenous lands as a result of the overall migratory flux into Roraima. Further agricultural projects or spontaneous invasions of colonists may, in the future, amplify this tendency, especially after the huge forest fires of 1998 burned the eastern limits of the Yanomami land in Roraima. Finally, in the last decade, various infra-structure projects have been conceived which would affect the Yanomami territory, including roads linking Boa Vista to the military bases of Projeto Calha Norte located along the Venezuelan border, construction of a hydroelectric dam on the middle Mucajaí (UHE Paredão), etc.

Figure 4. The military airstrip at Auaris in 1994, with piles laid in preparation for the construction of a new barracks.

Figure 5. Catholic mission buildings at Xitei, in the Parima highlands.

Ethnobotany and the Yanomami

Since the first anthropological studies were conducted amongst the Yanomami in the 1950s, mainly by German anthropologists such as O. Zerries and H. Becher, a great deal has been written about diverse aspects of their society and culture (e.g. Albert, 1985; Chagnon, 1968; Cocco, 1987; Colchester, 1982; Lizot, 1984; Ramos, 1995; Smole, 1976; Zerries and Schuster, 1974). In terms of ethnobotanical information, however, the literature is still relatively weak

Figure 6. Yanomami hunter in the forest.

when compared with the available quantity and diversity of anthropological information. The Yanomami use of plant-based hallucinogens has been addressed in considerable detail (e.g. Schultes and Holmstedt, 1968), as has their unusually detailed knowledge of edible fungi (Fidalgo and Prance, 1976; Prance, 1984) and their use of palms (Anderson, 1978). Prance (1972a) published general ethnobotanical data resulting from his extensive experience among the Yanomami in Brazil, and Fuentes (1980), Lizot (1984) and Finkers (1986) produced what are to date the most comprehensive studies of the Yanomami's use and knowledge of plants in general (based on fieldwork carried out in Venezuela). Valero (1984), in her description of her time as a 'captive' among the Yanomami, also makes numerous references to plant use. Nonetheless there remains a great deal to be learnt about their complex relationship with the plant world, as shown by recent revelations of their rich knowledge of medicinal plants (Milliken and Albert, 1996, 1997a).

This monograph presents the results of ethnobotanical fieldwork conducted between 1993 and 1995 among the eastern Yanomami of Brazil, principally with the 'lowland' Yanomami of *Watoriki* village (Demini, State of Amazonas) and to a lesser extent with the 'highland' or 'upland' Yanomami communities of Xitei (State of Roraima). An additional visit was made to communities of the Balawaú area in the State of Amazonas (Fig. 1). The principal aim of this fieldwork was the compilation of the Yanomami's fast-vanishing knowledge of medicinal plants, in a form potentially useful to future generations of Yanomami and also available to the medical personnel currently working amongst them, in the hope that they might be encouraged to value this knowledge as an alternative to the wholesale application of 'Western' medicines. During this fieldwork, however, information was gathered on many other aspects of the use of plants by the Yanomami. The results of the first surveys of medicinal plants, and of a detailed survey of the construction materials used in the *Watoriki* communal round-house (*yano*), were initially published elsewhere (Milliken and Albert, 1996, 1997a, 1997b). This monograph summarizes the complete body of ethnobotanical information that was gathered during this fieldwork, and places it in the context of previously published data on the Yanomami and on Amazonian ethnobotany in general.

The study sites

The village *yano* (communal round-house) of *Watoriki*, where the majority of the original data presented in this book were collected, housed a population of 89 at the time of the 1993 study (110 in 1999), of whom 46% were children under the age of ten. It is situated close to Km 211 on the abandoned Perimetral Norte (BR210) highway. Close to the village, on a section of the abandoned highway, there is an airstrip (1°30'48"N, 62°49'22"W, alt. 154m a.s.l.), and an indigenous post (PIN Demini) run by Yanomami spokesman Davi Kopenawa on behalf of the Fundação Nacional de Índio (FUNAI, the state administration of indigenous affairs which

Figure 7. The village of *Watoriki* in 1996, with a recently planted garden in the background.

succeeded SPI in 1967). There is also a medical post run by the Comissão Pró Yanomami (CCPY), a Brazilian NGO responsible for the medical support for the Yanomami in that sector of their territory. The people of *Watoriki* (Demini) village moved down from their territories on the upper Lobo d'Almada (a tributary of the Rio Catrimani) into the Mapulaú basin (a tributary of the Rio Demini) in the early 1970s. After decimation by an unidentified epidemic in 1973, contact with a FUNAI settlement on the Mapulaú in 1974–75, and then a further lethal epidemic of measles in 1977, they progressively approached the Demini post where they settled into permanent contact at the beginning of the 1980s. This migration was the last move of the *Watoriki* people in the context of the general expansion of the Yanomami of the highlands of the Serra Parima into the lowlands of the Rio Branco tributaries (see above).

The village is situated in dense lowland evergreen rainforest. The forest is diverse and mixed, with a fairly typical structure and composition for the region (see Huber, 1995; Milliken and Ratter, 1989), including species characteristic of both the Amazon and the Guianas. *Watoriki* village lies at the edge of the massif which divides the Amazon and Orinoco rivers; the largely flat Rio Branco basin extends to the south and southeast and the hills of the Serra Parima rise to the north and northwest. The country rock of the area consists of metamorphics of the lower Pre-Cambrian Guiana Complex, and the soils are largely clayey, dystrophic red-yellow latosols (RADAMBRASIL, 1975), but there are patches of conspicuously sandy soils in the vicinity of the village.

The second site, Xitei, lies in the highland part of the Yanomami territory, in the headwaters of the Parima river. At Xitei there is a Catholic mission (Fig. 5), a FUNAI post, an airstrip (2°36'40"N, 63°52'28"W, alt. 620m a.s.l.), and one Yanomami community (*Watatasi*). The mission and the FUNAI post were opened in the early 1990s after the end of the gold-rush, Xitei having been a tin-ore *garimpo* between 1988 and 1989. In the surrounding area there are 20 other Yanomami communities (754 persons), at one of which (*Kuai u*) fieldwork was also carried out. The third site, Balawaú, lies at the foot of the highlands to the north-west of *Watoriki* village. At Balawaú a health post with an airstrip (1°48'13"N, 63°47'55"W, alt. 60m a.s.l.) has recently been set up by the CCPY (1993), close to which one of the nine communities of the area has settled (242 persons).

All of these sites are in dense evergreen tropical rainforest. The vegetation at Balawaú is essentially similar to that in the vicinity of *Watoriki*, although some elements of the vegetation found there, e.g. *Piper francovilleanum* and *Tabernaemontana macrocalyx*, are absent at *Watoriki* but are common in the highland forests. The vegetation in the Xitei region is fairly typical of the submontane *terra firme* forests of the Serra Parima. These have been described in detail by Huber *et al.* (1984), who set the 600 m contour as the approximate lower level of these submontane forests and recognized them as distinct from the forests at lower altitudes. However, although there are noticeable differences between these submontane forests and the forests in the *Watoriki* region, including a greater richness in the epiphyte flora, a slightly lower canopy, and the inclusion of species typical of the Guayana Highland (e.g. *Psammisia guianensis*), they also include many of the common elements of the flora of the neighbouring lowlands such as *Aspidosperma nitidum*, *Bauhinia guianensis* and *Cedrelinga catenaeformis*.

Presentation of plant names in this monograph

Data from the literature, particularly when the authors have restricted themselves to Yanomami plant names, have only been included where those names can be translated with a reasonable degree of confidence into scientific names. In order to keep the text relatively readable, Yanomami plant names, plant families and authorities are presented separately (Appendix 1). Scientific names have generally been updated to conform with currently accepted taxonomy. However, in some cases it has been judged that this would cause unnecessary confusion, and hinder comparisons with other ethnobotanical data. For this reason the well known palm species *Maximiliana maripa* and *Jessenia bataua*, for example, have been retained in spite of the fact that Henderson (1995), in his rather drastic treatment of Amazon palm taxonomy, has incorporated them into *Attalea* and *Oenocarpus* respectively. The need to make such decisions about whether or not to ignore the 'latest' taxonomic thinking raises the question of whether taxonomists should exercise a greater degree of pragmatism or caution when dealing with the nomenclature of useful and well-established species. The highly subjective and variable nature of the conclusions arrived at by such specialists leaves enormous scope for confusion (see Kahn, 1997).

RESEARCH METHODS

Discussions of the properties and uses of plants were conducted directly in the Yanomami language in 1993 (July–August) and 1994 (July–August). In June 1995 (at Xitei and Balawaú), when the fluent Yanomami speaker (BA) was not present, the information was all recorded on audio tape and later transcribed and translated in the field by him. Information was provided by the Yanomami collaborators in three principal contexts:

- During interviews conducted with the specific purpose of identifying plant species employed for particular purposes.
- During interviews conducted with the purpose of constructing a Yanomami lexicon.
- Spontaneously, either in the forest (when a useful plant was encountered) or in the village.

The names of plants were recorded using the current standard orthography for Yanomami language in Brazil[2]. In almost all cases (except where the species was identifiable without doubt), ethnobotanical data were supported by herbarium voucher specimens, initially preserved in 70% ethanol (Schweinfurth technique). These were either collected simultaneously with the recording of the data, or subsequently. Fertile specimens have been lodged at the herbaria of

---

[2] The orthography used to transcribe the Yanomami words used in this article (in bold italics) generally follow the standard IPA usage. The **Yanomae** dialect spoken at **Watoriki** village has seven vowels (i, e, i̵, ë, a, u, o) and thirteen consonants (p, t, k, t^h, hw, s, x, h, r, m, n, w, y). Of the vowels, only two are represented by non-IPA standard symbols: i̵ refers to a close/tense, high central unrounded vowel (which has no corresponding sound in English) and ë which refers to a mid central vowel (similar to schwa /q/ vowel in the English word *cut*). Of the consonants, t^h refers to an aspirated, valveolar stop (similar to the initial consonant in the English word *tip*), **hw** refers to a labialized, voiceless glottal fricative (similar to the initial sound in the word *where* for some dialects of English), and **x**, following Portuguese orthography, refers to a voiceless palatal fricative /ʃ/ (similar to the initial sound in the English word *ship*). The palatal semiconsonant/semivowel /j/ is represented by **y**, according to common American phonetic usage (see Albert and Goodwin Gomez, 1997).

the Instituto Nacional de Pesquisas da Amazônia at Manaus (INPA), the Royal Botanic Gardens, Kew (K), the Museu Integrado de Roraima in Boa Vista (MIRR) and the New York Botanical Garden (NY), and a full set including sterile voucher specimens is maintained at Kew. Collection of specimens was carried out in the forest with Yanomami informants (principally Justino and Antonio at *Watoriki*), and plant identifications and usage data were subsequently checked by consensus in the village. Data were double-checked with as many people as possible (from two to several), depending upon availability. Finally, systematic discussions of the entire data set were held with a group of adult men[3] in the village (using only the plant names as prompts), both in 1993 and in 1994. Identification of a plant species was only recorded when a consensus agreement was reached. On the rare occasions that un-corroborated use data were specifically denied during group discussion, they were rejected.

PLANTS AND THE YANOMAMI

Variation in Yanomami plant knowledge

There exists considerable variation, both qualitative and quantitative, in Yanomami knowledge of the properties and identifications of plants, both at the level of the individual and at the level of the community. This was particularly evident for medicinal plant knowledge, which formed the main focus of the fieldwork in this study. In other words, within one community there were people who knew a great deal more about medicinal plants than others of the same age group in the community, and the medicinal plants known to one person tended to differ considerably from those known to another. These differences were also found in other aspects of Yanomami ethnobotany, although to a lesser extent. The number and range of species known to individuals as being useful for a particular purpose will inevitably be influenced by a number of factors, one being the transmission span for that information from their parents, i.e. the number of years they lived before their parents died. Other potentially important factors (depending on the nature of the knowledge) include their sex and age (see Phillips and Gentry, 1993b), their past migrations, and in some cases the origins of their parents (one of whom may have come from another village or area).

The overall level of plant knowledge present within any particular community also varies, depending on the degree of influence by outsiders (either indigenous or not), on the number of surviving members of the older generations, and on the length of time which they have lived in the region

---

[3] The majority of the primary data presented in this paper were collected with male informants, as a consequence of the concentration of the research on medicinal plant information. This was largely provided (at *Watoriki*) by older adult males, the older women of the community (who traditionally were primarily responsible for using these medicines) having all died (see 'Medicinal plants'). As a result, there is possibly an under-representation of the ethnobotanical knowledge of women, which may differ in some aspects from that of the men.

they inhabit. There may also be considerable qualitative differences between the plants used in one community and those used in another, as was found between **Watoriki** and Xitei in the case of the medicinal species (Milliken and Albert, 1997a). One of the reasons for these differences is the variation in forest composition across the Yanomami area, particularly with the altitudinal changes between the highlands and the lowlands. However, in some cases a plant (e.g. **apūru uhi**) was found to occur in both regions but only to be recognized as a medicinal species in one of them. In other cases, plants were found to occur and to be recognized as medicinal species in both regions, but for different medicinal purposes, (e.g. **ixoa hi**). This serves to illustrate a variation which is probably widespread amongst Yanomami communities and which extends beyond the knowledge of medicinal plants.

The names used for specific plants can also vary from one part of their territory to another, reflecting the linguistic differences which exist between the sub-groups of the tribe. Even within a single community, more than one name may be used for a single plant species. The plant known at **Watoriki** village as **weyeri hi**, for example (*Protium fimbriatum*), is known as **mani hi** in the Xitei region, and at **Watoriki** *Aspidosperma nitidum* is referred to both as **hura sihi** and **poo hētʰo hi**. Conversely, in a few cases a single Yanomami name is used for two or more plant species – even within a single community – although the properties of those plants may not be the same. In the Xitei region, for example, the name **xitokoma hi** is used for both *Swartzia schomburgkii* (Leguminosae) and *Chimarrhis* sp. (Rubiaceae)[4]. These types of nomenclatural variation are not uncommon amongst the indigenous communities of the Amazon (see Bennett and Andrade, 1991).

Nominal classifiers in Yanomami plant names

Lizot (1996) has listed many of the classifiers used with plant names among the western Yanomami, and Fuentes (1980) has provided a list of the words used for plant anatomy and morphology. No detailed analysis of the nomenclatural and classificatory systems used by the Yanomami for plants and their component parts was made during this study, and such an analysis would require a new in-depth combined linguistic and botanical effort. However, one interesting linguistic aspect of Yanomami plant nomenclature which can be dealt with here is the short, usually one or two syllable, word which follows the noun in the plant name and indicates the category or class to which the plant belongs, or, if this word is also a plant/body part designator, the part of the plant being described. Among plant names a limited number of these classifiers regularly recur and are reasonably predictable for species for which they are perceptually and/or functionally significant. The presence of these classifiers may indicate some general property or salient attribute of the

---

[4] These trees look superficially similar (they both have lobed trunks), but they are not related to each other in the botanical sense and possess very different useful properties.

referent of the noun, such as shape or other physical characteristic, although the defining characteristic may not be easily discernible. Such words, known as nominal (or noun) classifiers, are common in Amazonian languages. The use of classifiers in plant names can serve to group together species which, for the Yanomami, possess similar physical characteristics or practical uses, and thus shed light on their perception of plant relationships.

The occurrence of nominal classifiers in the Yanomami languages is not restricted to plant nomenclature but extends into other domains, e.g. to indicate different animal species (*naki* for bees and wasps and *kiki* for snakes). Dozens of these so-called 'characteristic classifiers' (Borgman, 1990) have been suggested for the various Yanomami languages. The comments made here are limited to those classifiers which were found to occur in plant nomenclature collected from speakers of the **Yanomae** dialect in **Watoriki** village. While only the most common classifiers for plant names will be discussed, the total number of possible noun classes is much greater.

By far the most frequently occurring classifier for plant names is *hi*, with 40% or 235 occurrences out of a total database of 584 entries. This classifier, which is found in the majority of the names for trees, is transparently derived from the generic term for tree, *huu tihi*. It has a large number of related, derived classifiers, such as *ahi, uhi, usihi, unahi, nahi, xihi, axihi, kohi*, etc. These derived classifiers may relate to particular characteristics. *Usihi*, for example, is derived from the word *usi* (soft), and the genera to which it is applied are notable for their soft wood (e.g. *kahu usihi*, *Cecropia*; *ara usihi*, *Croton*). The other most frequently occurring classifiers used specifically for plant names are *tʰotʰo, si* and *hanaki*, each with 43 or 44 occurrences in our data set. The three general classifiers *a* (generic singular), *ki* (generic plural) and *kiki* (plural for clustered objects), although not specific to plants, are also common, with 38, 34 and 28 occurrences respectively.

Of the 49 plant names which identify vines, creepers and lianas, 44 include the characteristic classifier *tʰotʰo*. The other five, mainly fish-poison vines, occur with the classifier *āthe*. A category broadly comprising the Palmae, Musaceae and Marantaceae (i.e. Monocotyledons) can generally be identified by the presence of the characteristic classifier *si* (plural *siki*). The classifier *hanaki* (*hana* 'leaf' + *ki* 'plural') occurs in 44 names, generally for plants made up mainly of leaves (e.g. *Heliconia bihai*, *Peperomia* spp., *Psychotria ulviformis*). It can also be used to refer specifically to the leaves of other plants, especially when those leaves are attributed with medicinal, magical, or ornamental properties.

The collective classifier *kiki* (not to be confused with the *koko* classifier for manioc tubers) identifies a category of plants possessing tubers or small rhizomes which are edible or used for medicinal or magical purposes. It is also used to indicate the edible tuberous part of vines (e.g. several varieties of sweet potatoes) in which the plant name otherwise occurs with the classifier *tʰotʰo*. In addition, the collective *kiki* is one of four classifiers which categorize fruits. Consistent with its usage for certain animals and objects which occur in groups or bands (*warë kiki* is a band of peccaries), *kiki* identifies fruits which occur in clusters such as bananas (*paixima kiki*) or peach palm fruits (*raxa kiki*).

Fruits with large kernels and thin mesocarps are generally identified by the characteristic classifier *maki*, which is possibly derived from *maa ma* (stone). These include fruits of a number of palm trees, such as *mai maki* (*açai* palm fruits) and *hoko maki* (*bacaba* palm fruits). Fruits which are very small are categorized by the characteristic classifier *moki*, which is also the word for seeds. Finally, the fourth category of fruits is indicated simply by the addition of the plural *ki* to the tree/plant name, such as *aso asiki* (fruits of the *aso asihi* tree).

Twenty-eight tree/plant names in the data set contain only the generic classifer *a*, which can be used with most Yanomami nouns to denote the singular. Although this classifier is frequently used with names of animals or other objects, it is less commonly used with plants. It seems probable that its use in plant names reflects the fact that these species are associated with specific substances or objects, taking the classifier *a* along with the names of those substances or objects (e.g. *xīkī a*, from whose bark *xīkī* cords are made). In the cases of *yākoana a*, and *horoma a*, used to produce *yākoana* snuff and *horoma* snuff-tubes respectively, the plants may also be referred to as *yākoana hi* and *horoma si*. The latter form tends to be used in the context of discussion of botanical differentiation, whereas the *a* form is used in discussion of the technological uses of the plants, i.e. reference to the plant as the object it will become. A number of the plants in the *a* class (e.g. *xõwa a*) are used as ingredients in substances for sorcery or magic.

Yanomami horticulture

Swidden clearings

Like most Amazonian indigenous peoples and indeed most forest peoples, the Yanomami practise swidden horticulture. This was not analysed or explored in detail during the present study but has been described and discussed in depth by Hames (1983b), Harris (1971), Lizot (1980) and Smole (1976, 1989). Small patches of forest are felled and burned (Fig. 8), and crops are planted and harvested for the few years in which the group remains in the area. The principal cultivated plants (*hutu theri thëki*) grown in these swidden clearings or gardens are discussed under 'Plants as food'.

Ecological factors and the choice of swidden sites

The choice of optimal swidden sites is inevitably affected by the technology available for their clearing and cultivation, as has been discussed by Colchester (1984) and Lizot (1980). The relative inefficiency of the stone axe in comparison with the steel axe in the felling of large trees has been analysed and discussed by many authors, including Carneiro (1979a, 1979b) in the Yanomami context. Some have concluded that the felling of tall forest for swidden cultivation with stone axes would never have been put in practice effectively. This would have been compensated for by the application of alternative or additional clearing techniques and by careful choice of location (Denevan, 1992).

Figure 8. A Yanomami garden, with paw-paw tree in the foreground, cassava behind and bananas and plantains in the distance. Edible white bracket fungi are beginning to develop on the felled trunks.

This was certainly the case for the Yanomami. According to the oldest informants [interviewed by BA], the undergrowth was cleared with digging-sticks (*sihema*) and small machete-like palm wood blades (*poo*) (Fig. 9), and the smaller trees (*oxe tihi* or *hiya tihi*) were felled with hatchets (see below for descriptions and discussion of Yanomami tools). Larger trees were killed by ring-barking and/or by burning their bases. However, these would have been scarce in the plot since the future garden area would have been selected where the vegetation was relatively easy to clear using the available technology. Thus, although the Yanomami were quite capable of efficient swidden horticulture prior to the arrival of steel tools, these did allow them to cultivate in a greater variety of forest types (and therefore areas), thus presumably facilitating their migrations and the expansion of their range.

Figure 9. Tools used in the past for the clearing of gardens: the machete-like **poo** (70 cm), made from palm wood, and the hatchet-like **haowatima**, made from a fragment of steel set in a stick.

Figure 10. Felling a *Jacaratia digitata* tree with a steel machete.

According to Smole (1976), the optimal areas for cultivation were regarded as being those which were dominated by stands of musaceous plants such as *Heliconia* and *Phenakospermum*, and Finkers (1986) notes that the presence of *Ceiba pentandra*, *Martiodendron* sp., *Micropholis* sp. and *Theobroma bicolor* are taken to indicate that the land is good for cultivating bananas, whereas the presence of *Clathrotropis macrocarpa* indicates that it is unsuitable for bananas but good for cassava and yams. Cocco (1987), who also provides an overall description of gardening practices, lists *Heliconia* and *Calathea* spp. (among others) as positive indicators, and according to Valero (1984) the

Table 1
Some plant species occuring in forest patches formerly used for swidden cultivation in the highlands by the *Watoriki* Yanomami

Family	Species	Habit	Yanomami name
Anacardiaceae	*Spondias mombin*	Tree	**pirima ahi tʰotʰo**
Begoniaceae	*Begonia humilis*	Herb	**morimoripë**
Bignoniaceae	*Jacaranda copaia*	Tree	**xitopari hi**
Caricaceae	*Jacaratia digitata*	Tree	**rihuwari si**
Commelinaceae	*Buforrestia candolleana*	Herb	Name unknown
Dioscoreaceae	*Dioscorea piperifolia*	Vine	**kaxetema**
Euphorbiaceae	*Margaritaria nobilis*	Tree	**yãpi mamohi**
Gramineae	*Olyra* sp.	Herb	**purunama asi**
	Orthoclada laxa	Herb	**tʰomi koxi kiki**
	Pharus virescens	Herb	**xikirima hanaki**
	Indet.	Herb	**wayapapë**
Guttiferae	*Rheedia macrophylla*	Tree	**kotaki axihi**
Heliconiaceae	*Heliconia bihai*	Herb	**irokoma hanaki**
Lauraceae	Indet.	Tree	**pokara mamohi**
Leguminosae	*Acacia polyphylla*	Tree	**kayihi una tihi**
	Inga edulis	Tree	**krepu uhi**
	Piptadenia sp.	Tree	**ërama kiki**
	Indet.	Tree	**oko hi**
Marantaceae	*Calathea* aff. *mansonis*	Herb	**makerema asi**
	Calathea sp.	Herb	**saruma asi**
	Ischnosiphon arouma	Herb	**mokuruma si**
Meliaceae	*Guarea guidonia*	Tree	**mãri hi**
	Trichilia pleeana	Tree	**hwatho ahi**
	Trichilia sp.	Tree	**akanaxi ahi**
Piperaceae	*Piper* sp.	Herb	**mahekoma hi**
	Pothomorphe peltata	Shrub	**mahekoma hi**
Sterculiaceae	*Theobroma bicolor*	Tree	**himara amohi**
	Theobroma cacao	Tree	**poro unahi**
Tiliaceae	*Apeiba membranacea*	Tree	**taitai ahi**
Urticaceae	*Urera* sp(p).	Shrub	**apina siki**
Zingiberaceae	*Costus* spp.	Herb	**naxuruma aki**

group with whom she lived used the presence of wild yams on yellow soil to indicate good land. She describes the clearing of patches of low vegetation with hatchets when there were no steel tools available, and also mentions the trial planting of tobacco as a means of determining the suitability of an area for cultivation.

These observations correlate to some extent with data collected by the present authors. A list of some of the plant species which were said to have been common on cultivable land (**hutu kana**), as recognized by the **Watoriki** people when they were living in the highlands, was recorded during the present study and in previous investigations by BA (see Table 1). Lucas, one of the older **Watoriki** men, described the ideal site as having an undergrowth rich in *Heliconia*, a soft fine-grained black or grey soil with few stones and no large roots, and a vegetation which is easy to crush and burn. He pointed out, however, that the limit of these patches is situated near the **Xopata u** river (a small tributary of the Couto de Magalhães[5] in the upper Mucajaí basin) – a limit which may have presented a barrier to Yanomami migration prior to the introduction of more effective cutting tools.

A large proportion of the species listed in Table 1 are herbaceous, probably representing the composition of vegetation types which are relatively easy to clear using the traditional technology as has been discussed. However, it is also possible that some of them are indicators of soil chemistry or fertility. The cocoa tree *Theobroma cacao*, for example, has been described by Smyth (1975) as 'exceptionally demanding in its soil requirements', with a root system 'particularly sensitive to factors within the soil which impede root growth and a poor ability to obtain nutrients and moisture when these are in short supply'. Thus it presumably serves as a sound indicator of good soils. Similarly, the emphasis on an abundance of musaceous plants (*Heliconia* and *Phenakospermum*) in the undergrowth may represent more than just ground which is easily cleared, but also a suitability for the cultivation of their relative the banana.

The use of plants as indicators of soil fertility or suitability for cultivation is known from studies among other Amazonian groups (e.g. Milliken *et al.*, 1992) and is probably a common phenomenon. In many parts of the Amazon, black anthropogenic soils known locally as *terra preta do índio*, rich in phosphorus, calcium and carbon and usually containing numerous pot-sherds, are the preferred sites for swidden horticulture. These soils are thought to mark the locations of previous villages, and to indicate repeated occupation over very long periods (Balée, 1989; Smith, 1980). Such soils, however, are not common in the Yanomami territory. Although some Yanomami groups (including the **Watoriki** people) did, during their migrations from the highlands, occupy old swidden sites left by earlier migrating Yanomami groups or by extinct tribes formerly living in the lowlands, the soils at these sites do not approach *terra preta* levels.

---

[5] This seems to be roughly the boundary between the highlands (**horepë a**) and the lowlands (**yari a**).

Plants as food

Cultivated food plants

At the time of study, the most important food plants (**wamotima siki**) grown in the swidden clearings by the Yanomami at **Watoriki** village were cassava (*Manihot esculenta*), which is the staple starch source as it is for most Amazonian peoples (see Plate 8b), and bananas and plantains (*Musa* spp.). Other significant crops included pawpaw (*Carica papaya*), maize (*Zea mays*), sugar cane (*Saccharum officinarum*), peach-palm (*Bactris gasipaes*), yam (*Dioscorea trifida*), sweet potato (*Ipomoea batatas*), and taro (*Xanthosoma* sp.).

Figure 11. Planting in a recently established garden. The plant in the foreground is *Clibadium sylvestre*, a cultivated source of fish poison.

Figure 12. Paw-paw trees and young peach-palm.

In the immediate vicinity of the village, fruit trees and shrubs had been planted including pimenta (*Capsicum frutescens*), ice-cream bean (*Inga edulis*), avocado (*Persea americana*), cashew (*Anacardium occidentale*), lime (*Citrus aurantiifolia*), orange (*C. sinensis*), mango (*Mangifera indica*), pineapple (*Ananas comosus*), Barbados cherry (*Malpighia punicifolia*), coconut (*Cocos nucifera*) and guava (*Psidium guajava*), the last eight of which are recent introductions[6]. Indian shot (*Canna indica*) is also grown as a crop in some Yanomami villages, and Finkers (1986) records the cultivation of *Calathea altissima* for food.

The Yanomami maintain a broad genetic base for their crop plants. Finkers (1986), for example, recorded 14 banana cultivars among the Yanomami with whom he worked in Venezuela, and although no systematic study of crop cultivars was made at **Watoriki**, linguistic studies revealed the names of 12 banana varieties (including plantains), nine cassava (including four no longer grown by the group), four sweet potato (including one no longer grown), four yam, three pawpaw, three peach palm and two sugar cane. Another (unsystematic) survey of crop cultivars, carried out [by BA] at the village of Toototobi in 1981, yielded a similar diversity of varieties (see Table 2) although the degree of overlap between the names recorded in these two communities, apart from those of the bananas and plantains, is not great.

The diet of the Yanomami, like any other aspect of their material culture, has inevitably been adapted over time to suit changes in their lifestyle and resources and to exploit new opportunities. Perhaps the most significant of these adaptations is the overall shift from what is presumed to have been an almost total reliance on wild food plants towards a strong dependence of cultivated species[7] (see Colchester, 1984). However, within the cultivated species themselves there have been significant changes. Maize and plantains, for example, played a more important part in their diet in the past than they do today, and most of the old maize cultivars have been lost and replaced by mass-produced varieties from modern agriculture (**napë yonomo si** or 'white man's corn'). According to the older inhabitants of **Watoriki**, their remote ancestors used to celebrate a **reahu** (the inter-village alliance and funerary ritual) with maize or with *Micrandra* seeds, as recounted in many myths, instead of with the cassava which is used today or the boiled plantains which were usually preferred before contact. Bananas and plantains, however, were themselves introduced from the Old World at some point, and their acquisition may have precipitated the intensification of Yanomami horticultural practices. It has been suggested that this acquisition could have brought about a dramatic increase in the Yanomami population (Harris, 1974), with a number of possible concomitant consequences such as village fission and migration to the lowlands (Colchester,

[6] An unusually high number of recently introduced fruit tree species was found at **Watoriki**, most of which were brought in by one individual.

[7] Lizot (1978) estimated agriculture (horticulture) as contributing over 73% of the food (by weight), 77% of the energy and 26% of the protein in the diet of the communities which he studied in Venezuela.

Table 2
Crop cultivars recorded at Toototobi village in 1981

Crop	Cultivar name	Descriptor/Notes
Cassava (bitter)	*aurima koko**	white (from ***napë*****)
	hayokoari koko	mythical tapir-like animal
	hutuwisasi koko	capuchin monkey tail
	puuxirima koko	short
	wakërima koko	red
	yanaema koko*	ant
Cassava (sweet)	*raparima koko*	long
	uxirima koko	black (from ***napë*****)
	yãhihirima koko*	sticky
Yam	*aurima aki**	white
	keteterima aki	sweet
	uxirima aki*	black/deep blue
Taro	*aurima kiki*	white
	hwaxoma kiki	
	ixaro kiki	red-tailed japím bird
	krokohorima kiki	grey
	uxirima kiki	black/deep blue
Sweet potato	*ãkõtarima kiki**	
	õsimõhõrima kiki	
	wakërima kiki	red
Peach palm	*raxa araasi kiki*	yellow macaw?
	raxa paxo kiki	spider monkey
	raxa apia kiki	*Micropholis* tree
	raxa roa kiki	
	raxa wakarai kiki*	light/pale
Banana (sweet)	*manito si**	(from ***Watoriki***)
	paixima si*	
	rëprare	
	rokoa si*	
	rokoma si [F]	
	uxipirima si* [F]	(from Catrimani)
	xiahima si*	
	yarimina si	
Banana (plantain)	*kõhitarima si*	
	koraha si yai	
	koraha yawere si*	(from ***Watoriki***)
	monakarima si*	
	monapirima si [F]	
	pareama si* [F]	
	piriko si	(from ***napë*****)
	yatopra si	
Maize	*uxirima moki**	black grains
	hrare moki	orange grains
	aurima moki	white grains (from ***napë*****)
Sugar cane	*uxirima uki**	black (from highlands)
	mayëpë uki	toucan (from highlands)
	aurima uki	white (from highlands)
Anatto	*manokorima si**	bold
	irepërima si*	white-haired

[F]   Corresponding names recorded by Finkers (1986)
*   Corresponding names recorded at ***Watoriki***
** The term ***napë*** is applied to all non-Yanomami people, but refers here specifically to 'white
     people'.

Figure 13. Stretching the tubular *tipiti* basket to squeeze the juice from grated cassava.

1984). This intensification of horticulture would also have stimulated a significant increase in sedentariness among Yanomami communities.

The older inhabitants of **Watoriki** claimed that their ancestors possessed only a few varieties of bitter cassava, and that the sweet varieties (***witatirima si***) were acquired from other Indian groups and white people at the time when the first pieces of metal were arriving in their communities, i.e. around the turn of the century. The high-producing bitter varieties (***kõaimirima si***) now constitute the principal starch source at **Watoriki**, and the use of this resource has certainly been facilitated by the borrowing of Carib technology (the tubular *tipiti* basket) for squeezing the poisonous juices from the grated tubers (Fig. 13), replacing their more basic traditional techniques (see 'Plants for tools, implements and miscellaneous uses'). This borrowing probably also occurred around the turn of the century, through indirect contact with the Ye'kuana and other neighbouring groups now extinct. In the past, cassava pulp was baked on a piece of broken pottery or on fresh *Phenakospermum* leaves over a small fire, or simply rolled into balls and cooked in the embers wrapped in leaves. These comparatively rudimentary cooking techniques, combined with the equally rudimentary methods used for grating the tubers[8], are indicative of a relatively late acquisition of cassava by the Yanomami.

Despite the fact that they now practise a sophisticated system of agriculture (see Hames, 1983b, Harris, 1971, Lizot, 1980, Smole, 1989), the Yanomami are traditionally a trekking society (Colchester, 1984, Good, 1995). They used to spend between one-third and one-half of the year away from their villages, travelling long distances through the forest, staying for short periods in temporary camps and relying heavily on wild food sources. A variety of wild seeds and tubers replace cassava and bananas as starch sources during these periods. Lizot (1984, 1997) discusses the varying seasonal availabilities of these wild food species and their relative importance in the diet, and Zerries and Schuster (1974) provide a calendar of resource use. The hunting and gathering activities engaged in during trekking are described in detail by Fuentes (1980).

Wild food plants

The wild starchy seeds eaten by the Yanomami include those of *Clathrotropis macrocarpa, Dioclea* aff. *malacocarpa, Inga paraensis, Inga* sp. (**kāi hi**), *Micrandra rossiana* and *Plukenetia abutaefolia* (Fig. 14). Neither *M. rossiana* nor *D.* aff. *malacocarpa* are found near **Watoriki** village, although these and **kāi hi** are common in the highlands. All of the above seeds are to some degree toxic, and must be processed by grating, soaking, washing, boiling etc. (the process depending upon the species), in much the same way that bitter cassava is processed in order to rid it of its cyanogenic compounds. These processes are described in detail by Fuentes (1980), including the preparation of the toxic seeds of a *Pouteria* species which are so bitter that six hours of boiling are required before they can be consumed. Lizot (1984) mentions an unidentified species which, if eaten excessively by those who are unused to it, can cause haemorrhaging.

Similar uses of poisonous seeds as starch sources are recorded among many other Amazonian tribes. The seeds of *Clathrotropis macrocarpa*, for example, are

---

[8] Slabs of rough granite were used in the past for grating cassava (see photograph in Zerries and Schuster, 1974). Other traditional techniques and tools are discussed under 'Plants for tools, implements and miscellaneous uses'.

Figure 14. The split fruit of *Plukenetia abutaefolia*, showing the large edible seeds.

used in the same way by the Waimiri Atroari (Milliken *et al.*, 1992), *Plukenetia* seeds are eaten in Peru (Altschul, 1973), and in Equador the seeds of *Inga ilta* are eaten by the Quichua after boiling for 20–30 minutes (Pennington, 1997). The toxicity of *Inga* seeds is referred to by the Waorani of Ecuador, who claim that more than five ingested seeds of certain species can provoke severe vomiting (Davis and Yost, 1983).

Not all of the wild seeds eaten by the Yanomami are toxic, but most need some kind of preparation before they can be eaten. Notable exceptions are those of the Brazil nut *Bertholletia excelsa* and of the *Caryocar pallidum*[9]. The seeds of the false banana *Phenakospermum guyannense* are roasted in the embers of the fire before consumption. The use of this species for food does not appear to be common in the Amazon, but has been recorded among the Andoque in Colombia (La Rotta, 1983). The wild edible starchy roots and rhizomes, which are also cooked before they are eaten, include *Calathea* aff.

[9] Fuentes (1980) identified this species. No collection was made during the present study, although consumption of **xoxo** seeds was observed at Xitei. One collection made at **Watoriki** under the same name (**xoxomo hi**), which was said to be eaten in the highlands, was identified as *C. glabrum* (which has been reported to be eaten throughout its range), so there remains some confusion over the identification of this plant. It may be that both species are eaten but are known by the same Yanomami name. It may also be that there is little significant difference between these species: Prance and Silva (1973), in their revision of the genus *Caryocar*, stated that "the three species *C. glabrum*, *C. montanum* and *C. pallidum* form a group of closely related species which are often difficult to distinguish".

Figure 15. Edible tubers of *Calathea* aff. *mansonis*.

mansonis (Fig. 15), a wild variety of the cultivated yam *Dioscorea trifida, D. piperifolia, Dracontium asperum* and a wild sweet potato (**hōkōmo urihit^heri a**). According to Fuentes (1980), the roots of *Iryanthera ulei* are likewise cooked and eaten[10].

Leaves play very little part in the Yanomami diet, although those of a number of the plants which they cultivate are eaten elsewhere in the world (see Milliken *et al.*, 1992), and there are doubtless many wild species whose leaves could also be eaten if desired. In fact, in stark contrast to the ethnobotany of Southeast Asia, there appear to be few records of the consumption of leaves of any kind amongst Amazonian indigenous populations, although Barfod and Kvist (1996) have recorded consumption of fern fronds and other 'vegetables' amongst the Cayapas of the coastal rain forests of Ecuador, where at least 24 species are eaten.

Fruits, however, are important in the diet of the Yanomami, and a massive variety is eaten. Some of the more obviously important genera are *Inga, Hymenaea, Pourouma, Pouteria* and *Theobroma*. The larger palms are particularly important in terms of dietary contribution, including *Astrocaryum aculeatum, Euterpe precatoria, Jessenia bataua, Mauritia flexuosa, Maximiliana maripa, Oenocarpus bacaba* and several other smaller species (see Anderson, 1978). The

---

[10] The use of these *Iryanthera* roots as food (Fuentes, 1980) is rather surprising. This is probably an erroneous record.

Figure 16. Fallen fruits of *Theobroma bicolor*, a species much appreciated by the Yanomami.

mesocarps of these palms are rich in oils, proteins and vitamins, and constitute an important dietary supplement, particularly when game is scarce. The mesocarp of the fruits of *J. bataua*, for example, provides proteins which are said to be comparable in terms of amino acid content with the proteins from high quality animal flesh, and 40% higher in 'biological value' than soya proteins (Balick and Gershoff, 1981). The fruits of *M. flexuosa* contain very high levels of vitamin A (Aguiar *et al.*, 1980).

Most of the palm species also furnish nutritious and pleasant-tasting palm-hearts which are occasionally harvested; of the 20 species listed by Anderson (1978) only those of *Geonoma*, *Mauritia*, *Orbignya* and *Socratea* were not eaten.

Fungi and flowers

Only seven species of edible fungi were collected at **Watoriki** village in 1994, although the names of a further six species were given. The collected species were *Favolus brasiliensis*, *F. spathulatus*, *Filoboletus gracilis*, *Lentinus tephroleucus*, *Marasmius cubensis*, *Pleurotus flabellatus* and *Polyporus grammocephalus*. Generally speaking, fungi are little eaten by the indigenous peoples of Amazonia (Fidalgo, 1965). The Waimiri Atroari, for instance, claim to eat only one species, and that only very occasionally (Milliken *et al.*, 1992), and a number of other tribes have been recorded as eating only one or two species in situations of hunger (Fidalgo and Hirata, 1979, Prance pers. comm.). The Yanomami, however, are something of an exception in this respect, as was

Figure 17. *Filoboletus gracilis*, an edible fungus.

demonstrated by Prance (1984), who recorded 21 species of edible fungi from a single village and reported these to play a significant role in their diet[11]. These fungi are generally wrapped in leaves and roasted in the fire before eating. Most of the edible species grow on rotting logs, which are abundant in swidden clearings and thus provide a useful 'second crop' from abandoned sites. Of the species collected during the present study, only *Favolus brasiliensis* was recorded by Prance (see Appendix III), who found it at each of his three study sites and cited it as the most popular species, but he did record other species of *Lentinus*, *Pleurotus* and *Polyporus*. Fungi were not seen to be eaten in significant quantities at **Watoriki**, and this may be because there was a comparative abundance of meat available there at the time, making dietary supplements unnecessary. In the highlands, where large game is generally more scarce, the use of edible fungi, as well as other minor protein sources such as shrimps, termites, caterpillars, frogs, toads, etc., appears to be considerably greater.

The flowers of at least one tree species, **nai hi**, were reported to have been eaten by the elders of **Watoriki** when living in the highlands, and although it was not collected the same information has been recorded by Fuentes (1980), who identified it as *Manilkara bidentata*. Flowers do not appear to be eaten commonly by Amazonian indigenous peoples, and it is interesting that Barfod

---

[11] According to Prance it is the women who are most knowledgeable about edible fungi, and this was borne out by the present study.

and Kvist (1996), in their study of the ethnobotany of coastal Ecuador, found that one tribe (the Cayapas) were exceptional in this respect, consuming the flowers or buds of at least ten species whilst their neighbours the Colorados and the Coiaquers ate none.

Food plant diversity

Many of the plants used as sources of food by the Yanomami are also employed for other purposes (medicine, construction, technology etc.). An overall list of wild food plants of the Yanomami (excluding fungi), including the data from the present study and from Fuentes (1980) and Anderson (1978)[12], is presented in Table 3. The most important families in terms of numbers of species eaten (Palmae, Leguminosae, Moraceae, Sapotaceae) correspond very closely with those registered by Phillips and Gentry (1993a) in a forest hectare in Peru[13]. Clearly not all of the species will be found or eaten in any one area, since altitudinal differences preclude the occurrence of some of them in certain parts of the territory. *Bertholletia excelsa*, *Caryocar villosum* and *Phenakospermum guyannense*, for example, and some of the edible palms, are not encountered in the highland regions of the Yanomami territory (Huber *et al.*, 1984). Fuentes (1980) records 60 wild food plants from his study in Venezuela, Lizot (1984) records 65, and Anderson (1978) records 17 edible palms from the Toototobi region about 90–100 km to the NW of *Watoriki* village. Of the 60 species recorded by Fuentes, 22 were regularly eaten in the study area and regarded as significant in the diet, including the genera *Anacardium*, *Astronium*, *Bertholletia*, *Brosimum*, *Caryocar*, *Clathrotropis*, *Dacryodes*, *Dioscorea*, *Hymenaea*, *Inga*, *Jessenia*, *Mauritia*, *Maximiliana*, *Micropholis*, *Oenocarpus*, *Phenakospermum*, *Pouteria*, *Pseudolmedia*, *Sorocea* and *Theobroma*. According to Huber *et al.* (1984), the western Yanomami (Yanõmamɨ) of the Parima highlands recognize more than ten 'varieties' of edible *Dacryodes* (*mõra mahi*). Combining the available data, the overall total of almost 120 species is impressive, but still probably represents less than half of the edible wild species actually used by the Yanomami[14].

It is interesting to compare these figures with those obtained from other ethnobotanical studies with Amazonian Indians, even though one cannot draw solid conclusions since the existing data on edible plants of the Amazon are still very incomplete. Balée (1994), in his comprehensive account of Ka'apor ethnobotany, cites 179 edible non-domesticates of which 13 species are categorized as 'primary food sources'. This list includes nine species of *Eugenia*, 21 *Inga*, 14 *Pouteria* and nine *Protium*. Davis and Yost (1983) recorded

[12] These data were collected from Toototobi, to the northwest of *Watoriki*. In the 1970s the *Watoriki* people lived on the Mapulaú river, much closer to Toototobi than they do now.

[13] These families, listed by descending 'family use value', were the Palmae, Leguminosae (Mimosoideae), Sapotaceae and Moraceae.

[14] The names of a further 27 wild food plants, which have not yet been identified, were recorded at *Watoriki* during linguistic studies, and there are doubtless a great many more.

44 (regularly eaten) wild food plants with the Waorani of Ecuador, Glenboski (1983) recorded 73 wild food plants with the Tukuna (Tikuna) of Colombia, Boom (1987) recorded 75 with the Chácobo of Bolivia, and La Rotta (1983, 1988) recorded 38 and 46 with the Andoque and Miraña, respectively. Our list includes plants from a range of altitudes (and therefore forest compositions) covered by the Yanomami area, which may to some extent explain its size. However, it is likely that the discrepancies in numbers of edible wild species recorded among Amazonian indigenous groups are more a reflection of the differing levels of study carried out, and the differing emphases within those studies, than on the true usage levels. This subject is discussed further by Balée (1994).

Table 3
Some wild forest food plants eaten by the Yanomami

A = Recorded by Anderson (1978) but not this study; F = Recorded by Fuentes (1980) but not this study, FI = recorded by Finkers (1986) but not this study; M = this study

N.B. the *x* used in the Yanomami plant names recorded at **Watoriki** (following the currently accepted orthography in Brazil) corresponds to the **sh** used in earlier publications by Fuentes and others.

Species	Yanomami name*	Part eaten	Source
ANACARDIACEAE			
Anacardium giganteum	*oru xihi*	Fleshy pedicel	M
Astronium lecointei	*ōpōni*	Fruit	F
Spondias mombin	*pirima ahi tʰotʰo*	Fruit	M
ANNONACEAE			
Fusaea longifolia	*hwapoma hi*	Fruit	M
APOCYNACEAE			
Couma macrocarpa	*operema axi hi*	Fruit	M
Tabernaemontana sananho	*tʰoru hwātemo hi*	Fruit	M
ARACEAE			
Dracontium asperum vel aff.	*kātārā āsi*	Rhizome (boiled)	M
BOMBACACEAE			
Catostema sp.	*shēpre kë u fi*	Fruit	F
BORAGINACEAE			
Cordia nodosa	*xiho xihi*	Fruit	M
BROMELIACEAE			
Aechmea sp.	*reorima siki*	Fruit	FI
Bromelia tarapotina	*sitio kë siki*	Fruit	F
BURSERACEAE			
Dacryodes peruviana	*mōra mahi*	Fruit	M
Protium polybotryum	*hwaximo kohosi hi*	Fruit (aril)	M
Tetragastris altissima	*xoko hwaka hi*	Fruit (aril)	M
CARICACEAE			
Jacaratia digitata	*rihuwari si*	Fruit	M
CARYOCARACEAE			
Caryocar pallidum	*xoxomo hi*	Seed	M
Caryocar villosum	*ruapa hi*	Fruit (mesocarp)	
		Seed (roasted)	M

31

Species	Yanomami name*	Part eaten	Source
CHRYSOBALANACEAE			
Couepia caryophylloides	*wāro uhi*	Fruit	M
Couepia sp.	*hiha amo hi*	Fruit	M
COMPOSITAE			
Wulffia baccata	*totori mamoki*	Seed (semi-ripe)	FI
CONVOLVULACEAE			
Ipomoea cf. *batatas*	*hōkōmo urihiťheri a*	Tuber (cooked)	M
CUCURBITACEAE			
Posadaea sphaerocarpa	*pora axi*	Seed (roasted)	M
DIOSCOREACEAE			
Dioscorea piperifolia	*kaxetema*	Tuber (boiled)	M
Dioscorea trifida	*wāha urihiťheri a*	Tuber (boiled)	M
Dioscorea cf. *triphylla*	*yayomi*	Tuber (boiled)	F
Dioscorea sp.	*rai*	Tuber (boiled)	M
EUPHORBIACEAE			
Micrandra rossiana	*momo kë fi*	Seed (processed)	F
Plukenetia abutaefolia	*xōrahe ťhoťho*	Seed (processed)	M
GUTTIFERAE			
Rheedia benthamiana	*oruhe hi*	Fruit	M
Rheedia macrophylla	*kotaki axihi*	Fruit	M
HELICONIACEAE			
Heliconia bihai	*irokoma hanaki*	Young furled leaf	M
Heliconia sp.	*shiramiamoki*	Seed (cooked)	FI
LAURACEAE			
*Persea americana***	*ahuāri hi*	Fruit	M
LECYTHIDACEAE			
Bertholletia excelsa	*hawari hi*	Seed	M
LEGUMINOSAE			
Clathrotropis macrocarpa	*wapo kohi*	Seed (processed)	M
Dialium guianense	*paro koxihi*	Fruit	M
Dioclea aff. *malacocarpa*	*kuapara ťhoťho*	Seed (processed)	M
Hymenaea courbaril	*hatohato koxihi*	Fruit	M
Hymenaea parvifolia	*arō kohi*	Fruit	M
Inga acreana	*pahi hi*	Fruit (aril)	M
Inga acuminata	*rīa moxiririma hi*	Fruit (aril)	M
Inga alba	*moxima hi*	Fruit (aril)	M
Inga edulis	*krepu uhi*	Fruit (aril)	M
Inga myriantha	*ōkiōkirimi*	Fruit (aril)	F
Inga nobilis	*wākama*	Fruit (aril)	F
Inga paraensis vel aff.	*toxa hi*	Fruit (aril), Seed (processed)	M
Inga pezizifera ***	*kāi hi*	Fruit (aril), Seed (processed)	M
Inga pilosula	*poataa hi*	Fruit (aril)	M
Inga sarmentosa	*pooko hi*	Fruit (aril)	M
Inga scabriuscula	*kasha nafi*	Fruit (aril)	F
MALPIGHIACEAE			
Byrsonima aerugo	*atama asi*	Fruit	M
MARANTACEAE			
Calathea cf. *majestica*	*pisa asi*	Seed (roasted)	M
Calathea aff. *mansonis*	*makerema asi*	Tuber (boiled)	M
Calathea sp.	*kuma maki*	Tuber (cooked)	M
Monotagma sp.	*kopari hanaki*	Young furled leaf	M

Species	Yanomami name*	Part eaten	Source
MELASTOMATACEAE			
Bellucia grossularioides	**pitima hi**	Fruit	M
MENISPERMACEAE			
Abuta grandifolia	**itahimosi hi**	Fruit	M
Cissampelos pareira****	**kashitemi**	Root (cooked)	FI
MORACEAE			
Bagassa guianensis	**hapakara hi**	Fruit	M
Brosimum cf. *alicastrum*	**rimo mofi**	Fruit	F
Brosimum guianense	**ihuru makasi hi**	Fruit	M
Cecropia sciadophylla	**kahu usihi**	Fruit	M
Clarisia ilicifolia	**hihō unahi**	Fruit	M
Helicostylis tomentosa	**xopa hi**	Fruit	M
Maquira calophylla	**teria hi**	Fruit	M
Perebea angustifolia	**rëxë hi**	Fruit	M
Perebea guianensis	**peesisima hi**	Fruit	M
Pourouma bicolor ssp.	**ōema ahi**	Fruit	M
Pourouma melinonii	**waraka ahi**	Fruit	M
Pseudolmedia laevigata	**hayi hi**	Fruit, Seed (processed)	M
Pseudolmedia laevis	**aso asihi**	Fruit, Seed (processed)	M
Sorocea muriculata ssp. *uaupensis*	**yipi hanaki**	Fruit	M
MUSACEAE			
Phenakospermum guyannense	**ruru asi**	Seed (roasted)	M
MYRISTICACEAE			
Iryanthera ulei	**fäifäiyomi nat^{h}a**	Root (cooked)	
	faiyomi		F
MYRTACEAE			
Eugenia flavescens vel aff.	**pore hi**	Fruit	M
Myrcia sp.	**totori mamo hi**	Fruit	M
PALMAE			
Astrocaryum aculeatum	**ëri si**	Fruit, Palm heart	M
Astrocaryum gynacanthum	**xoomo si**	Fruit (roasted)	M
Astrocaryum murumuru	**maha si**	Fruit, Palm heart	M
Bactris monticola	**mokamo si**	Fruit	M
Bactris sp.	**yarimo si**	Fruit, Palm heart	A
Bactris sp.	**yoroa si**	Fruit, Palm heart	A
Bactris sp.	**yoyoma si**	Fruit, Palm heart	A
Euterpe precatoria	**maima si**	Fruit, Palm heart	M
Geonoma deversa	**warama si**	Fruit	M
Jessenia bataua	**koanari si**	Fruit, Palm heart	M
Jessenia polycarpa	**yei ki si**	Fruit	F
Mauritia aculeata	**torea si**	Fruit	A
Mauritia flexuosa	**rioko si**	Fruit	M
Mauritiella armata	**kuai si**	Fruit	M
Maximiliana maripa	**okorasi si**	Fruit, Palm heart	M
Oenocarpus bacaba	**hoko si**	Fruit, Palm heart	M
Orbignya spectabilis	**kunuana si**	Fruit	A
Scheelea martiana	**yoi si**	Fruit, Palm heart	A
Socratea exorrhiza	**manaka si**	Fruit	M
PASSIFLORACEAE			
Passiflora coccinea	**naxuruma t^{h}ot^{h}o**	Fruit	M
Passiflora vitifolia	**oroshomi**	Fruit	FI
QUIINACEAE			
Lacunaria jenmani	**paari makasi hi**	Fruit	M
Quiina florida vel aff.	**naxuruma ahi**	Fruit	M

Species	Yanomami name*	Part eaten	Source
RHAMNACEAE			
Zizyphus cinnamomum	*mirama asihi*	Fruit	M
RUBIACEAE			
Duroia eriopila	*hera xihi*	Fruit	M
Isertia hypoleuca	*yāpi uhi*	Fruit	M
SAPINDACEAE			
Paullinia cf. *pinnata*	*shuu funaki*	Fruit	F
Talisia cf pedicellaris	*mako ahi*	Fruit	M
SAPOTACEAE			
Ecclinusa guianensis	*yāre hi*	Fruit	M
Manilkara bidentata	*nai hi*	Flower	F
Manilkara huberi	*xaraka ahi*	Fruit	M
Micropholis melinoniana	*apia hi*	Fruit	M
Pouteria caimito	*paxo hwātemo hi*	Fruit	M
Pouteria cladantha	*hōrōmana hi*	Fruit	M
Pouteria hispida	*yāwaxi hi*	Fruit	M
Pouteria sp.	*poxe mamokasi hi*	Fruit	M
Pradosia surinamensis	*werihisi hi*	Fruit	M
STERCULIACEAE			
Herrania lemniscata	*xuhuturi unahi*	Fruit	M
Theobroma bicolor	*himara amohi*	Fruit pulp, Seed	M
Theobroma cacao	*poro unahi*	Fruit	M
Theobroma microcarpum	*prōō masihi*	Fruit	M
Theobroma subincanum	*waiporo unahi*	Fruit	M
THEOPHRASTACEAE			
Clavija lancifolia	*horikima hi*	Fruit	M

* Names recorded by Fuentes and Finkers are written using a differing orthography from those recorded during the present study.
** It is not clear whether avocado trees occur naturally in the forest or have escaped from cultivation by other (extinct ?) peoples.
*** The specimen was collected in the lowlands whereas the plant was said to occur mainly in the highlands. The Yanomami name may therefore refer to another species of *Inga*.
**** This may have been misidentified. The Yanomami name refers to *Dioscorea piperifolia* at **Watoriki**, a wild yam which is superficially similar in the sterile state.

Borderline species

Not all of the edible species recorded in this study are regularly eaten or highly appreciated foods. The fruits of *Astrocaryum murumuru*, for instance, were said by one informant to be edible and by another to be inedible. When pressed on the matter, it became clear that it was very much a matter of personal taste. The fruits of *Brosimum guianense* may be eaten but are not particularly relished (one man dismissed them as parrot food), and those of *Geonoma deversa* and *Socratea exorrhixa* are generally regarded as famine food, for use when there are no better alternatives. Similarly, although the 'hearts' (the young furled leaves) of *Heliconia bihai*, *Monotagma* sp. and probably several other related species may sometimes be pulled out and eaten, it is only when food is scarce that they are likely to be eaten in earnest.

It was said that the fallen fruits of *Bagassa guianensis* must be eaten as soon as they reach the ground, or they rapidly become unpleasant. Some people will not eat these fruits even then, but regard them as fit only for game. It was also said that if a nursing mother eats too many of the fruits of *Theobroma bicolor*, her baby will contract oral thrush (a yeast infection), and that eating the fruits of *B. guianensis*, *Helicostylis tomentosa*, *Spondias mombin* and certain other forest trees can cause sub-dermal maggot infestations[15]. These fruits rapidly become infested with small maggots when lying on the forest floor, and an association could have been made with this fact. Although this connection might seem unlikely from a scientific standpoint, it is interesting that the Ka'apor also claim that the fruits of *B. guianensis* cause botfly infestations (Balée, 1994). According to Finkers (1986) the eating of excessive numbers of Brazil nuts is associated with allergies and skin inflammation, to which some people are more susceptible than others.

Plants for salt and water

Although the Yanomami generally eat their food unseasoned, vegetable salt is occasionally used. This is obtained from the ashes of the giant forest tree *Couratari guianensis* or from the burned stems of the cyclanthaceous climber *Asplundia* sp. The ashes of the latter species contain calcium and (to a lesser degree) magnesium and potassium salts (data from analysis of a specimen collected by BA). Salt may also be obtained from the ashes of *Sanchezia* sp. *Couratari* is also used as a source of salt by the Bora of Peru (Flores Paitán, 1987), and the Waimiri Atroari burn the bark of its relative the Brazil-nut tree *Bertholletia excelsa* for the same purpose. Another cyclanthaceous species, *Cyclanthus bipartitus*, is used for salt in Peru (Denevan and Treacy, 1987), and the Taiwanos add the ashes of the leaves of *Asplundia ponderosa* to the food of pregnant women, possibly as a nutritional supplement (Schultes and Raffauf, 1990). A considerable variety of other plants are used as sources of vegetable salt the world over.

Plants may also be used as sources of water, which is occasionally necessary when travelling by foot in the dry season. A number of large dilleniaceous forest lianas provide substantial quantities of clear potable drinking water when the lowest loops of their stems are cut, and at least one of these, *Pinzona coriacea*, is used for this purpose by the Yanomami. This practice is also a very common one in Amazonia.

Drug and stimulant plants

Hallucinogens

Details of the preparation and use of hallucinogenic snuffs by the Yanomami for shamanic purposes have been recorded in considerable detail by numerous authors (Biocca, 1979; Brewer-Carias and Steyermark, 1976; Chagnon *et al.*, 1970, 1971; Fuentes, 1980; Prance, 1970; Schultes and

---

[15] These maggots are apparently similar to the larvae of the botfly (*Dermatobia hominis*).

35

Figure 18. A shaman's equipment hanging from the rafters of the **yano**. The feathers and furs are for adornment, and the diagonal wooden tube is for administering the hallucinogenic snuff.

Holmstedt, 1968; Seitz, 1967) and will not be repeated here. The principal sources of these hallucinogens are the bark resins of *Virola* spp. and the seeds of *Anadenanthera peregrina*. These plants contain tryptamines (Agurell *et al.*, 1969), which are responsible for their psychoactive properties, and plants of both genera are used by other South American indigenous groups for the same purposes (Altschul, 1972; Schultes and Raffauf, 1990). *Anadenanthera* trees are sometimes planted in swidden clearings by the Yanomami (Finkers, 1986).

The snuff is administered by a second, who blows it down a long **horoma** tube (60–90 cm) traditionally made from the hollowed stem of *Iriartella setigera*, at one end of which a rounded pierced palm seed (*Maximiliana maripa*) is glued with resin in order to provide a seal for the nostril (Fig. 18). Although at **Watoriki** it is now more common for a piece of arrow cane (*Gynerium sagittatum*) to be used for this purpose, as recorded by Fuentes (1980) in Venezuela, *Iriartella* is particularly suitable for tube-making on account of its slender stem, soft inner pith and hard outer wood, and is employed by the Waorani of Ecuador for making blow-guns (Davis and Yost, 1983). The stems of *Olyra* sp., *Ischnosiphon* spp. or *Bactris monticola* can also serve for making these tubes.

At **Watoriki** village three distinct 'varieties' of *Virola* were collected[16]. These were **yākoana a** (WM1780) which is the most commonly used, **haare a** (WM1917) which was said to be the strongest and to have been responsible for occasional deaths[17], and **xioka a**[18] (WM1992) which was said to be weak and to cause irritation of the nasal passages. All of these varieties fall within the somewhat broad taxonomic boundaries of the species *Virola elongata* (*sens. lat.*), which includes *V. theiodora*. Fuentes (1980) also records the use of *V. sebifera* from his study area in Venezuela. *Anadenanthera peregrina*, which is said to be more powerful than *Virola*, does not occur in the environs of **Watoriki** village, but its seeds are sometimes acquired through trade with other Yanomami groups, who may travel long distances to gather them[19].

Several plants are used as admixtures to the *Virola* snuff to enhance its effect, including the dried and powdered leaves of *Justicia pectoralis* var. *stenophylla* (Fig. 19) and the ashes of the outer bark of *Duguetia lepidota* and *Elizabetha leiogyne*. The use of *Justicia* and *Elizabetha* (*E. princeps*, a closely related species) by the Yanomami is already well documented, and Fuentes (1980) mentions the use of a *Duguetia* species in the preparation of this drug. He also

[16] One informant at **Watoriki** mentioned a further tree, **xamaxamarema a**, which was used in the past as a source of weak hallucinogen. This may be a *Virola* species.

[17] These are likely to have been men unaccustomed to taking the drug in great quantity (unlike shamans). All men take some at the end of the **reahu** ceremonies (see Albert, 1985 for a detailed description and analysis of **reahu**).

[18] Literally 'anus **yākoana**'.

[19] In 1993 a group of **Watoriki** men, on a visit to Brasília to meet representatives of the government administration, collected bags of seeds of this species from trees growing in the sports compound of the University.

Figure 19. Cultivated *Justicia pectoralis* var. *stenophylla*, used as an admixture to *Virola* snuff, growing outside the *yano*.

cites the use of *Aphelandra?* sp., *Isertia parviflora, Paullinia?* sp. and *Urera caracasana* in the preparation of 'drugs', but no mention was made of these at **Watoriki**. The dried leaves of *Justicia pectoralis* are smoked in the form of cigarettes by the Wayãpi of French Guiana (Grenand *et al.*, 1987).

Tobacco and tobacco substitutes

The Yanomami commonly keep a wad of tobacco leaves in the lower lip, which acts not only as a stimulant but also, by raising the pH of the saliva, seems to help prevent caries. The highly prized tobacco plants (*Nicotiana tabacum*) are planted in swidden clearings and around the **yano**. Fresh leaves (or moistened dried leaves) are softened in the fire and mixed with ashes of certain types of firewood (Fig. 20) before being rolled into an appropriately sized wad. Suitable ashes are produced from trees of the genera *Acacia, Aspidosperma, Capirona, Duguetia, Duroia, Elizabetha, Fusaea, Inga, Picramnia, Rheedia, Rinorea, Sagotia, Trichilia,* and doubtless several others. These may have been recommended on account of alkalinity of their ashes, as this is likely to enhance the effect of the tobacco, in the same way that lime is chewed with coca (*Erythroxylum*) leaves in other parts of South America. Schultes and Raffauf (1990) and Prance (1972b) describe the use of several plants as additives to tobacco snuff among the tribes of the west and northwest Amazon, including the powdered dried leaves of *Inga* and *Cephaelis*, and the ashes of *Geonoma, Hyospathe, Scheelea* and *Theobroma*.

At times when demand for tobacco leaves exceeds supply (not an uncommon situation in some areas), or when the Yanomami are travelling far from their villages, other plants may be used as tobacco substitutes, employed in roughly the same manner[20]. These include the rhizomes of *Zingiber officinale*, the roots of *Piper bartlingianum*, the inner bark of *Annona ambotay* and the leaves of *Besleria laxiflora, Erechtites hieraciifolia, Gossypium barbadense* and *Miconia lateriflora*. The leaves of a bush called **yaraka hanaki** (not identified) are also used. At Balawaú it was also said that the roots of *Piper francovilleanum* and *P. demeraranum* will serve this purpose. It is interesting to note that the Panare Indians of Venezuela chew the leaves of *Piper piscatorium* in the same manner as tobacco (Boom, 1990), and the Kulina of Peru use the dried and pulverized leaves and roots of *Piper interitum* as a substitute for tobacco snuff. Furthermore, several species of *Piper* are employed as stimulants and hallucinogens in the Peruvian Amazon and on the coast of Ecuador (Barfod and Kvist, 1996, Kvist and Holm-Nielsen, 1987). Others are used in Melanesia, e.g. by the Yali highlanders of West Papua (WM pers. obs.). The Gesneriaceae are not commonly used as stimulants, but the Siona-Secoyas do employ one member of this family (*Columnea picta*) in this manner, smoking its leaves in place of tobacco (Schultes and Raffauf, 1990).

---

[20] These substitutes tend to produce a stinging sensation in the mouth similar to that of tobacco, although whether they possess similar stimulatory qualities is not clear.

Figure 20. Mixing fresh tobacco leaves with hot ashes in the hearth.

Smoke from burning the dried outer bark of *Zollernia paraensis*, which peels off naturally in large slabs, is used to calm crying babies. The bark is burned beneath their hammocks, and it was said that the smoke can also produce a narcotic effect on adults. Likewise, if banana soup is prepared in a trough made from the wood of *Licaria aurea*, this may produce a slight narcotic effect on the drinker.

Plants employed in hunting and fishing

Weapons

The principal hunting weapon employed by the Yanomami is the bow and arrow. The bows, which are up to 2m long, are generally made from the trunks of palm trees, principally *Bactris gasipaes* and *Oenocarpus bacaba*, or (rarely) from the wood of a large forest *Swartzia*. One informant at **Watoriki** also mentioned the use of a "wide-trunked" palm called **kripi si** (possibly *Iriartea deltoidea*) and another (leguminous?) tree called **hixa ahi** for this purpose. Fuentes (1980) cites the use of a *Geonoma* species, presumably for rather small bows. According to Anderson (1978), *Socratea exorrhiza* and *Jessenia bataua* may also be employed for making bows, and although the wood of the cultivated *B. gasipaes* is said to be the strongest, it is often left standing on account of its other role as a highly valued fruit tree. Its wood is used to make spears and blow-guns by the Waorani of Ecuador (Davis and Yost, 1983) and is also used for hunting bows elsewhere in Brazil (Frechione *et al.*, 1989).

The bow was traditionally shaped and smoothed with the incisor of a collared peccary **poxe naki** (*Tayassu tajacu*), and the lip of the shell of a giant land snail **sinokoma aka** (*Melagalobulimus oblongatus*), with a final polish with the asperous leaves of *Pourouma bicolor* ssp. *digitata*. The last of these is still in use, but machetes and knives are now more commonly used for shaping and smoothing. Bows are strung with stout cords, now generally made from fibres of the silk-grass **yāma asiki** (*Ananas* sp.)[21], which may be strengthened and coloured by treatment with the sticky red resin from the inner bark of *Inga alba*. In French Guiana the sticky red sap from the inner bark of *Licania macrophylla* is applied to bow-strings in a similar manner (Grenand and Prevost, 1994). As an alternative to silk-grass, which is a cultivated species and therefore not available on long hunting or trekking excursions, twisted strips of the fibrous inner bark of *Cecropia* spp. are used for the bow-string. These bark fibres were in fact used exclusively in the past, before the acquisition of silk-grass. Fuentes (1980) also mentions the use of *Ficus* bark in this capacity.

The exceptionally long arrows used by the Yanomami are made from the peduncles of the flowering arrow-cane *Gynerium sagittatum*[22], which is

---

[21] According to one informant at **Watoriki**, a 'variety' of this plant called **marokorima asiki** also serves for cord and string.

[22] Three varieties of this plant are recognized by the Yanomami at **Watoriki**: **xaraka si yai** (real arrowcane), **xuhutima si** (itchy arrowcane) and **napë si** (foreigner arrow cane).

Figure 21. Arrow cane (*Gynerium sagittatum*).

cultivated in swidden clearings. A wild cane, **hwaya si**, may also be used. Arrows are generally fletched with the wing-feathers of the curassow (*Crax* spp.), bound to the shaft with fine cotton thread (Fig. 22). The cotton binding may be dyed purple with an extract of the leaves of *Picramnia spruceana* or red

Figure 22. Block of *mai koko* resin and curassow-feathers, for arrow-making.

with the arils of *Bixa orellana*[23] (Plate 12a). Cotton is also generally used to bind the nock (the grooved piece of wood into which the bowstring fits), fore-shaft, etc., but tougher silk-grass fibres coated with the resin of *Symphonia globulifera* or *Moronobea* sp. (Fig. 22) are more often employed to fix the arrow-heads. The resin (*mai koko*)[24], which also serves as a glue, is collected as lumps which are exuded from wounds in the trees' trunks, and then purified by setting them alight and collecting the molten resin which drips from them as they burn. Lizot (1984) provides illustrations of the types of arrow-heads manufactured by the Yanomami.

The fine-grained wood of *Rinorea lindeniana* was specified as suitable for making the nock. Although *Rinorea* is probably only one of a number of genera potentially suitable for this apparently undemanding purpose, its specific use for nocks has also been recorded among other Yanomami groups (Finkers, 1986), reflecting either a particular suitability or a traditional practice. The arrow-heads used vary according to the types of prey sought. The simplest consists of a sharpened stick or fore-shaft, made from a suitably strong wood such as *Mouriri* spp., which is inserted in the shaft and bound with string. Fuentes (1980) cites *Mouriri myrtifolia* and *M. sagotiana* as appropriate species, and the latter is also used in this way by the Wayãpi of French Guiana (Grenand, 1980). For birds, fish and small game a slender fore-shaft is fitted

[23] This species is cultivated, and at least three varieties are recognised at *Watoriki.*

[24] This resin, mixed with soot, may also be used to dye the bow.

Figure 23. Yanomami arrow and the young spider monkey which it killed, showing the detail of the arrow's fletching.

with a small sharpened bone tip (from the radius of a monkey), which is bound on at an angle to give a barb as well as a point.

For heavy game and for warfare, broad lanceolate pieces of sharpened bamboo (*Guadua* spp.)[25] are bound directly to the arrow shaft. Several kinds of bamboo may be used for making these blades (***rahaka***), of which two species were collected at ***Watoriki*** and four varieties (species?) were recorded by Fuentes (1980). Cocco (1987) describes the decoration of these arrow-heads with dye from *Bixa orellana* fruits or *Picramnia* leaves. Branched sticks are used as arrow-heads for small birds. Certain saplings possess a branching pattern particularly suitable for this purpose (several branches radiating in close proximity to each other from the main stem), which also makes them ideal for use as 'swizzle-sticks' for mixing up plantain soup. Although specimens were not collected during the present study, these are probably *Mabea* species.

Poisoned and interchangeable arrow-heads

Monkeys are generally hunted with slender arrow-heads made from slivers of palm wood from *Iriartella setigera* or *Jessenia bataua*, coated with resin from the trunk of *Virola elongata*. This arrow poison is extracted from the bark with heat,

[25] The collective term for bamboos is ***rahaka aki***. These include ***si koropërima*** (rough-skinned), ***wana*** (quiver), ***si hrakarima*** (soft skin) and ***katana si***.

Figure 24. Arrow-making

smeared on the arrow-heads (which are scored so that they break off in the wound), and dried over the fire. Several coats are applied. The process is described in detail by Chagnon (1968). The wood of *Socratea exorrhiza* and *Bactris gasipaes* may also be used to make these poisoned arrow-heads (Anderson, 1978).

The **Watoriki** Yanomami also used to employ a curare arrow poison, prepared principally from the bark of a *Strychnos* vine, when they lived in the highlands, but this species is apparently not found in the area they currently inhabit. The preparation process is described by Lizot (1972). At Balawaú, a species of *Strychnos* known as **māokori** (*S.* cf. *hirsuta*[26]) was collected, and was said to be

[26] The specimen was sterile and hard to identify, and may represent *Strychnos guianensis.*

used as the basis for curare poison. Various additives (**māokori hwëri**) may be used to strengthen the curare, including the bark of *Tabernaemontana sananho* and the menispermaceous lianas *Abuta imene*, *A. grisebachii* and *Curarea candicans*. All of these lianas contain alkaloids, and all are reported elsewhere as used in curare preparation: *A. imene* is used by the Taiwanos and Makunas of Colombia, *A. grisebachii* by the Tukanos (Schultes and Raffauf, 1990) and *C. candicans* by the Warrau of Guyana (Krukoff and Barneby, 1970).

According to Fuentes (1980), *Clathrotropis macrocarpa*, *Geissospermum argenteum*, *Paragonia pyramidata*[27] and *Piper dilatatum* are also used by the Yanomami in the preparation of curare poisons. Finkers (1986), who also describes the manufacturing process in considerable detail, cites the use of *Strychnos guianensis* for the curare base with the addition of *Tabernaemontana* sp. and some of the species listed by Fuentes, whereas Lizot (1984) mentions the use of *Anomospermum* sp. and an unidentified Malpighiaceae as additives. Cocco (1987) reports the inclusion of *Virola elongata* and a plant called **yoawë sihi**, which is almost certainly *Coussapoa* sp. The use of *Piper* for this purpose is widespread (Schultes and Raffauf, 1990), and various bignoniaceous lianas (*Arrabidaea*, *Callichlamys*, *Distictella*, *Martinella* and *Schlegelia*) have been reported to be used as curare ingredients elsewhere (Gentry, 1992). Furthermore, although reference has not been found in the literature to use of the *Clathrotropis macrocarpa* and *G. argenteum* in this way, these species are known to be rich in alkaloids.

Arrow-heads may be interchangeable, the bindings left so that they can easily be undone when necessary. When the poisoned arrow-heads are not in use they are carried in a small quiver (**wana**) made from a short length of *Guadua* bamboo, capped with animal hide and often decorated with patterns burned into the sides (see Lizot, 1984, for an illustration). These are carried on the back, attached to a loop of string around the neck. A mixture of the various types of arrow-heads is often carried in these quivers, affording the hunter versatility without the need for carrying large numbers of specialized arrows. Similar bamboo quivers are used by the Waorani of Ecuador for carrying their poisoned blow-gun darts (Davis and Yost, 1983). Small children use the petiole fibres of *Jessenia bataua* to make miniature arrows (**ruhumasiki**) for practising, and to imitate a monkey arrow-head a spine from the trunk of *Bactris gasipaes* is pushed into the end of a piece of a rigid forest grass.

The techniques and materials described here for arrow-making are essentially a variation on a basic pattern found throughout the Amazon basin. The use of *Gynerium sagittatum* canes, *Bromelia* or *Ananas* (silk-grass) fibres, cotton thread, guttiferous resins, split bamboo, palm wood and fine-grained hardwoods is virtually universal in the region, although the size and form of the arrows and arrow-heads and the style of decoration, fletching, binding etc. inevitably vary considerably. A summary of these styles and techniques is given by Chiara (1987).

---

[27] This species is cited as **mamokori**, which is the name usually given to *Strychnos* by the western Yanomami.

Dogs, men and plants

The Yanomami use dogs for hunting, and certain plants such as *Clathrotropis macrocarpa* and *Zingiber officinale*[28] are used to improve their performance. This use of plants to stimulate hunting dogs, either by rubbing the crushed plants over the dogs' bodies or by forcing the dogs to eat or sniff them, is encountered world-wide and is common among Amazonian peoples (e.g. Balée, 1994; Milliken *et al.*, 1992; Schultes and Raffauf, 1990). It may be that the odours of these plants sensitize the noses of the dogs to the scent of the quarry they are seeking, or perhaps that they hide or disguise the scent of the dogs themselves from the quarry. Both of the aforementioned species are used to prepare the dogs for hunting agouti (*Dasyprocta* sp.), the bark of *C. macrocarpa* smelling strongly of one of the agouti's principal food sources (its seeds) and the rhizome of *Z. officinale* smelling of the animal itself.

At least one plant is used in a practical manner, albeit indirectly, to improve human hunting performance. According to Finkers (1986) hunters who have lost their skill cut down branches of *Tachigali paniculata* and cover themselves with the vicious stinging *Pseudomyrmex* ants which inhabit the leaf rachises. Likewise, the tails of dead scorpions[29] may be scratched along the arms for the same purpose (BA pers. obs.). The use of the pain of insect bites to stimulate hunters (both dogs and humans) is not particular to the Yanomami, but has been reported amongst many indigenous peoples, e.g. among certain tribes of the Guianas (Im Thurn, 1883). Plants are also used in more symbolic rituals which ensure success in the hunt, the ritual and the plants employed depending upon the game being sought. These are described in more detail under 'Plants in ritual, magic and myth'.

Game ecology

Yanomami hunters possess a phenomenal understanding of the ecology of their environment, which is vital to their success. The knowledge of which game animals feed from which plants, combined with a knowledge of the phenology and distribution of those plants, enables them to predict with a reasonable degree of accuracy where they are likely to encounter a particular animal at a particular time of year[30]. Lizot (1984) lists some of the plants eaten by categories of game animals, and a limited quantity of information of this type was recorded on an opportunistic basis at **Watoriki** village during the present study. However, a systematic analysis of this aspect of Yanomami ethnobotany, i.e. their knowledge of inter-specific relationships, would be a truly Herculean labour, and the selection of data presented in Table 4 represents only a tiny part of their knowledge.

---

[28] The Yali highlanders of West Papua use ginger (*Z. officinale*) in a similar manner to the Yanomami (WM pers. obs.).

[29] Scorpion (**Sihiri**) is a mythological hero with an enormous bow, whose shooting is unsurpassable.

[30] Hunting platforms (hides) are sometimes built in fruiting trees which attract small birds that are shot for ornament.

Table 4
Plant species eaten by game animals – a small sample of Yanomami ecological knowledge

Family	Species	Eaten by:
Anacardiaceae	*Anacardium giganteum*	Tortoises, tapir, curassow, monkeys, partridges
Anacardiaceae	*Spondias mombin*	Tortoises, tapir, peccaries, deer, monkeys, macaws
Apocynaceae	*Couma macrocarpa*	Tapir, nine-banded armadillo, tayra
Caryocaraceae	*Caryocar villosum*	Deer, agouti, tapir, squirrels, giant armadillo
Chrysobalanaceae	*Licania* aff. *heteromorpha*	Parrots and monkeys.
Euphorbiaceae	*Margaritaria nobilis*	Many bird species
Euphorbiaceae	*Sagotia racemosa*	Parrots and macaws
Lecythidaceae	*Couratari guianensis*	Macaws (immature fruits)
Lecythidaceae	*Eschweilera coriacea*	Macaws
Leguminosae	*Clathrotropis macrocarpa*	Agouti
Leguminosae	*Elizabetha leiogyne*	Howler monkey and macaws
Leguminosae	*Hymenaea parvifolia*	Monkeys, peccaries, macaws
Leguminosae	*Inga alba*	Parrots, macaws and monkeys
Leguminosae	*Inga paraensis*	Macaws and monkeys
Melastomataceae	*Bellucia grossularioides*	Tortoises and tapir
Meliaceae	*Guarea guidonia*	Toucans, toucanets and small birds
Meliaceae	*Trichilia pleeana*	Toucans and guan
Moraceae	*Bagassa guianensis*	Tortoises, tapir, deer, curassows, tinamous, trumpeters
Moraceae	*Brosimum guianense*	Parrots
Moraceae	*Brosimum lactescens*	Birds and agouti
Moraceae	*Clarisia ilicifolia*	Small birds
Moraceae	*Helicostylis tomentosa*	Tortoises, tapir, peccaries, curassows, partridges and tinamous
Moraceae	*Pourouma bicolor* ssp. *digitata*	Monkeys, macaws and parrots
Moraceae	*Pourouma minor*	Howler and spider monkeys
Moraceae	*Pourouma ovata*	Bats and parrots
Moraceae	*Pseudolmedia laevigata*	Toucans, parrots and other birds
Moraceae	*Pseudolmedia laevis*	Many bird species
Myristicaceae	*Iryanthera juruensis*	Monkeys, birds and collared peccaries.
Palmae	*Astrocaryum murumuru*	Tortoise, paca, peccary, tapir, agouti
Palmae	*Euterpe precatoria*	Toucan, toucanet and trumpeters
Sapotaceae	*Chrysophyllum argenteum*	Monkeys, parrots, macaws etc
Sapotaceae	*Manilkara huberi*	Monkeys, tapir and kinkajou
Sapotaceae	*Micropholis melinoniana*	Tapir, monkeys, deer, peccaries, paca, agouti
Sterculiaceae	*Theobroma subincanum*	Tayra and capuchin monkey

This knowledge, which is the result of intimate personal experience of the interactions between components of the rainforest ecosystem, combined to some extent with lore passed on from one generation to the next, extends far beyond the knowledge of the deportment of game animals. It constitutes a valuable facet of the ethnobotany of the Yanomami and other rainforest peoples which is too often ignored or under-valued, while scientifically trained

Figure 25. Fruits of the *tucumã* (*Astrocaryum aculeatum*) palm, partly eaten by agoutis.

biologists spend their lives trying to discover, through lengthy observation and experiment, information which is already common knowledge among the indigenous inhabitants. Data of this type are sparse and scattered among the ethnobotanical literature, making significant comparison difficult. It is clear, however, that there is some degree of correlation between the data recorded during the present study and those from others. For example, all of the species listed in Table 4 were also reported as food for game animals by Balée (1994)[31] and/or by Milliken *et al.* (1992). Although the animals cited were not always the same in the different studies, this is probably an indication of incomplete rather than contradictory data.

Insect ecology

Certain plant species, both wild and cultivated, are recognized as hosts for edible insect species, and are exploited accordingly. The **mãmoahupë** and **wayawayapë** caterpillars which infest the leaves of cassava plants are eaten by the Yanomami and according to some people the **yoroporipë** caterpillar which destroys tobacco plants used also to be eaten in the past. The trunk of *Jacaratia digitata* has a soft pithy centre which rapidly rots when felled, and becomes infested with large edible insect larvae. The cut end of the trunk may be blocked with a palisade of sticks to prevent armadillos from entering and

[31] Balée (1994) records 170 wild plant species as food sources for three or more species of game animals.

Figure 26. The rotting trunk of a felled *Jacaratia digitata* tree. The cut end has been blocked off with sticks in order to prevent armadillos from entering and eating the insect larvae which are developing inside.

consuming the larvae (Fig. 26). The fallen fruits (seeds) of *Maximiliana maripa* often contain the fat white larvae of a bruchid beetle[32], which are nutritious and taste of coconuts. The rotting trunks of *Mauritia flexuosa* and certain other palms are host to the large edible larvae (**mapiapë**) of the palm beetle *Rhynchosporos palmarum*, and other edible larvae (**kuyuhumapë**) are taken from the trunks of *Oenocarpus bacaba*. At a certain time of year in the highlands, edible **kaxapë** caterpillars are collected from the leaves of *Inga* sp., and others are taken from *Gouania frangulaefolia*, *Protium fimbriatum*, *Vismia guianensis* and many other wild plants in certain seasons. These caterpillars, which may be 'hunted' by actively searching the trees in which they are known to occur, can play a significant role in the diet at times when other sources of protein are sparse. The names of twelve species of edible caterpillars were collected during fieldwork at **Watoriki**, and Lizot (1984) mentions five species and lists 24 trees upon which they are known to feed. These include species of *Anacardium*, *Ceiba*, *Licania*, *Micropholis*, *Pourouma*, *Protium*, *Sterculia*, *Swartzia*, *Talisia*, *Theobroma*, *Touroulia* and *Virola*.

Honey, which is provided by several species of wild stingless forest bees, is highly appreciated by the Yanomami both for its taste and, in some cases, for its medicinal properties. The names of 28 of these species[33] were collected at

[32] These *gonga* larvae are used by the neighbouring Makuxi Indians as bait for *pacu* fish.

[33] Twenty of these were said to produce sweet honey, six 'sour' honey, and two were not defined.

Figure 27. The curved thorns of *Uncaria guianensis*, a common vine. These were used in the past as fish-hooks.

Watoriki, and people were found to possess a considerable knowledge of their diets and behaviour. It was said, for example, that *Couratari guianensis* is visited by *õi* bees, *Acacia polyphylla* by *õi* and **yoi** bees, and *Swartzia* sp. by **yoi** and **paxopoma** bees. Again, this subject has not been studied in a systematic manner but to do so would certainly yield fascinating information for entomologists, ecologists and pollination biologists.

Fishing

At *Watoriki* there are no large rivers in the immediate vicinity, and most fish are taken from small streams. The fish are small, and are taken either with hook and line or with plant-based fish poisons. Prior to the introduction of steel fish-hooks, the curved thorns of *Uncaria guianensis* (Fig. 27) were said to have been used for this purpose, tied on silk-grass or *Cecropia*-fibre lines. Carved pieces of the leg bone of the nine-banded armadillo (*Dasypus novemcinctus*) were also said to have served as hooks. The most efficient method of fishing these streams is by the poisoning of the water, an activity which is primarily undertaken by groups of women[34]. However, the indiscriminate nature of fish-poisoning can result in serious damage to fish populations if practised regularly in the same locations, and care is taken to leave sufficient recovery periods between successive poisonings of a stream.

[34] Some men do take part in fish poisoning, particularly in the beating and crushing of the vines.

Figure 28. Collecting the leaves of *Clibadium sylvestre.*

Nine species of fish-poison plants were recorded at **Watoriki** village, and a tenth was later reported from Toototobi (Maria-Inês Smiljanic-Borges, pers. comm.). The most commonly used plant is *Clibadium sylvestre* (Fig. 11), which is cultivated in the swidden clearings and around the **yano.** Large quantities of the leaves are pounded with poles in a hole in the ground in order to release the juices (Fig. 29), and the resulting pulp (mingled with soil) is transferred to the stream in a basket and agitated in the current. Further downstream, a line of women waits with loosely woven sieve-like baskets to scoop out the stunned fish as they pass by (Plate 8b), and other fish may be shot with arrows by accompanying men or children. The strength of the poison can be increased by the addition of the latex-rich bark of *Tabernaemontana sananho* or *T. angulata.* The latter species does not occur in the forest around **Watoriki** village, but is very common in the highland regions which the group used to inhabit. Another species which is grown in swidden clearings for the express purpose of fish-poisoning is *Phyllanthus brasiliensis,* and at Toototobi the leaves and fruits of the cultivated *Capsicum frutescens* are commonly used in combination with *Clibadium.*

Various wild forest plants are used for fish-poisoning, most of which are woody vines (lianas). These include *Banisteriopsis lucida, Lonchocarpus* cf. *chrysophyllus, L. utilis* and *Serjania grandifolia.* The bark of the giant forest tree

Figure 29. Pounding the leaves of *Clibadium sylvestre*.

Cedrelinga catenaeformis is also employed for the purpose. These woody plants are beaten over a log with a heavy stick before use, until sufficiently macerated to release their constituent chemicals.

The use of *Lonchocarpus* spp., *Clibadium sylvestre*, *Phyllanthus brasiliensis* and *Serjania grandiflora* as fish poisons is reasonably widespread in the northern Amazon and the Guianas. As a result of their regular and heavy use among the Ka'apor of the eastern Amazon, *Derris* (*Lonchocarpus*) vines are one of the first forest plant resources to become depleted in the vicinity of a village with advancing settlement age, stimulating a reliance on the cultivated *Clibadium sylvestre* (Balée, 1994). In the comprehensive survey of fish-poison plants composed by Acevedo-Rodríguez (1990), which includes 935 species, there are listed 33 species of *Lonchocarpus*, 10 species of *Clibadium*, 12 species of *Phyllanthus* and 57 species of *Serjania*. The activity of *Lonchocarpus* is due to the rotenone/lonchocarpin content of these lianas: powerful isoflavonoids which disrupt respiration of the fish at the mitochondrial level (Fukami *et al.*, 1967). *Serjania* acts through the surface-activity of the saponins found in the stem, which alter the surface tension of the water and thus block respiration at the gills. Saponins are encountered in abundance among the Sapindaceae, and many genera of this family are used for fish-poisoning. The active principles of *Clibadium* are ichthyothereol-type polyacetylenic acids (Czerson *et al.*, 1979) and those of *Phyllanthus* are probably lignans (Grenand *et al.*, 1987). *Capsicum frutescens* is used as an ingredient of arrow poison by the Wayãpi of French Guiana (Grenand *et al.*, 1987).

Acevedo-Rodríguez's list does not include *Banisteriopsis* species, and this may be the first record of their use as a fish poison. The mode of action is therefore unknown, but the genus is rich in bioactive compounds including alkaloids, polyphenols and saponins, and several species are known to be employed for their hallucinogenic properties[35] (see Schultes and Raffauf, 1990). There are also very few prior references to use of the genus *Tabernaemontana* for fish-poisoning. *Tabernaemontana muelleriana*, a species fairly closely related to those used by the Yanomami, is employed by the Bora of Peru (Denevan and Treacy, 1987), and another species is used in the Canary Islands (Acevedo-Rodríguez, 1990). The genus is known to contain a broad spectrum of alkaloids and other biologically active chemicals (including cardiac glycosides), and numerous species are attributed poisonous and medicinal properties (Schultes and Raffauf, 1990). Cardiac glycosides from other plant sources are used elsewhere for poisoning fish (Acevedo-Rodríguez, 1990).

Although there appears to be no prior record in the literature of the bark of *Cedrelinga catenaeformis* being used for fish-poisoning, a large leguminous tree (which was not identified) was recorded as used for this purpose among the Waimiri Atroari (Milliken *et al.*, 1992), and this has subsequently been named as *C. catenaeformis* (R. Miller pers. comm.) whose active principles have not been identified. In addition, Lizot (1984) reports the use of the unripe (saponin-rich) fruits of *Caryocar pallidum* for poisoning fishes by the Venezuelan Yanomami, which corresponds to the use of several other *Caryocar* species by the Maku, Tukano etc. in northwest Amazonia (Prance, 1990). According to Finkers (1986) the lowland western Yanomami in Venezuela also employ a species of *Inga* as an additive when preparing this poison.

Plants for body adornment

Dress

The traditional dress of the Yanomami men at **Watoriki** village consists of a simple cotton string worn around the waist, supporting the penis by the foreskin, and although today most people wear cotton shorts, this string is still generally worn underneath. According to Cocco (1987) *Cecropia* fibres also serve this purpose. The women wear a short apron (Fig. 30), woven from cotton and dyed red with *Bixa orellana*. Bunches of hollowed seeds of various kinds (**tiritirimopë**) and small shells (**sitipasipë**), which jangle pleasantly when agitated, are sometimes attached, together with pieces of colourful bird skins such as toucans, cotingas and tanagers. The women typically wear long pieces of the straw-like stem of *Andropogon bicornis* grass through the holes pierced in their lips and their nasal septa (Plate 11a). The same species is used in much the same way by the Bora of Peru (Denevan and Treacy, 1987). The slender cores of the young aerial roots of *Chrysochlamys weberbaueri* or, according to Fuentes (1980), the wood of *Rinorea riana,* may also be used for this purpose, and Cocco (1987) mentions

---

[35] This is the source of the drug known as *ayahuasca*.

Figure 30. Weaving a small apron from dyed cotton.

a species of *Protium* in this context[36]. Pieces of the stems of sturdy grasses or, in the case of the western Yanomami, of arrow-cane, are often worn in the pierced ear-lobes by men.

Necklaces are made from the seeds of *Renealmia alpinia*, threaded on a fine cotton or silk-grass thread while they are still damp and soft, and occasionally from the seeds of *Cardiospermum* sp. and *Coix lacryma-jobi*, both of which are commonly used for this purpose in South America. One informant at **Watoriki** also mentioned the use of the dried seeds of a small palm called **misikirima hanaki** (probably *Geonoma* sp.[37]) as beads, and Lizot (1984) reports the use of *Miconia* seeds likewise. Paradoxically the use of these seed necklaces is a recent introduction at **Watoriki**, inspired by contact with other indigenous groups. This has developed primarily as a means of providing exchange for glass beads brought in by outsiders, which constitute a rare and highly valued ornament preferred for personal use. In the past, the only necklaces used were made from pieces of aromatic wood or dried fragments of bulbs of certain plants (e.g. *Cyperus*), which were worn for specific purposes (see Plants in ritual, magic and myth). The wood of one of the species reported by Lizot (1984) to be used in this way, *Myroxylon balsamum*, also produces an aromatic resin which may be mixed with body paints.

[36] These species are apparently particularly suitable as their wood does not yellow after the bark has been stripped.

[37] This species, which was said to resemble *Geonoma baculifera*, is also used for thatching on the upper Catrimani river and its young red fruits are edible.

Body paints

The skin may be decorated with the black dye prepared from the unripe fruits of *Genipa americana*, purple from the leaves of *Picramnia spruceana*, (Fig. 31) or red from the arils of *Bixa orellana*[38]. The use of *Genipa* and *Bixa* as body paints is very widespread in the Amazon (Grenand and Prevost, 1994), and *Picramnia spruceana* is used as a purple dye by the Panare in Venezuela and the Waorani in Ecuador (Boom, 1990; Davis and Yost, 1983). According to Anderson (1978), dark-blue and black dyes obtained from the ripe fruits of *Jessenia bataua* and *Euterpe precatoria* are also used by the Yanomami for body painting. Fuentes (1980) mentions the use of *Genipa spruceana* for black paint (in the same way as *G. americana*), and the fruits of *Corynostylis arborea*[39] for violet, and Alès (1987), in her discussion of Yanomami body paints and perfumes, cites the use of *Renealmia* and *Hieronima* fruits. Cocco (1987) also describes a blue body paint produced from *Renealmia* fruits, and Prance (1972) recorded a purple dye made from a species of *Pourouma* by the Yanomami on the Upper Uraricoera. In addition, according to Lizot (1984) the leaves of *Picramnia macrostachya* are used by the Venezuelan Yanomami (in the same way as *P. spruceana*), and red and black dyes are prepared from exudates from the trunks of *Protium* sp. and *Tabernaemontana heterophylla*, respectively[40].

Decoration may take the form of complex patterns (see Cocco, 1987; Laudato and Laudato, 1984; Zerries and Schuster, 1974), particularly for ceremonial occasions, but often the body is dyed almost completely red with *Bixa orellana* (see Plate 12b). This may be mixed with the slightly fragrant oil from the trunk of *Copaifera* sp., which is not found near **Watoriki** village but is brought (traded) from elsewhere in small gourds, producing a dark red oily paint suitable for more intricate designs. This may be cooked to form a paste which is rolled into balls for later use. Lizot (1984) describes a similar process whereby *Inga* resin is cooked and mixed with *Bixa* arils and *Copaifera* oil to produce a body paint. The fine soot produced by burning the balls of resin (**warapa koko**) from *Hymenaea parvifolia* or from *Protium* spp. may be mixed with *Copaifera* oil and with *Bixa* paste to produce a black dye. This mixture is also employed by the Caribs of Guyana (Roth, 1924).

The people of **Watoriki** have ceased the skirmishing with neighbouring groups of Yanomami which was relatively common in the past. Before the men would go to raid another village, they would traditionally paint their bodies entirely black with a mixture of crushed charcoal and the sugary latex from the trunk of *Couma macrocarpa*. In 1993 the men at **Watoriki** village painted themselves black in this manner for the first time in many years (Fig 32),

---

[38] These dyes are sometimes used to colour the bodies (heads, legs) of dogs or other domesticated animals (monkeys etc.). They are also used on arrows, baskets, slings etc.

[39] The fruits of this species are hard and are not obviously coloured. This may be an erroneous record.

[40] The *Protium* and *Tabernaemontana* dyes are more likely to be used for dyeing artifacts than for body painting.

Figure 31. *Picramnia spruceana*, whose leaves produce a purple dye.

Figure 32. Body-painting with a mixture of crushed charcoal and *Couma macrocarpa* latex, after the 1993 Haximu massacre (see overleaf).

before setting out for Haximu where a massacre of Yanomami had been perpetrated by gunmen acting on behalf of a gold-miner (see Albert, 1994). However in this case they painted themselves and performed a war ritual as an act of animosity towards an outside aggressor, rather than another Yanomami group. The latex of *Couma macrocarpa* is also used to glue white down into the hair for ornamentation when visiting other villages, as is the sticky exudate from unripe bananas or the latex of *Himatanthus* sp.

Scents and blossoms

Women commonly wear bunches of leaves or flowers in their pierced ears and in cotton bands worn around their upper arms. Some of these plants are used to seduce men by their scent. The most frequently used leaves at **Watoriki** are those of *Justicia pectoralis*, which smell strongly of coumarin and are said to make men weak, the bright green soft young leaves of *Elizabetha* spp., and bundles of the epidermal layers stripped from young (furled) leaves of *Geonoma baculifera* (Fig. 55). The flowers of *Solandra grandiflora* and *Posoqueria latifolia* were seen to be used for this purpose, and it was said that the red flowers of *Sanchezia* sp., the white flowers of *Bauhinia guianensis* and the slender filament-like flowers of a spiny tree called **makuta asihi** (probably Bombacaceae) are also worn. *Solandra* flowers may also be rubbed on the body as a deodorant (or perfume). Doubtless many other plants are regarded as suitable for ornamentation, and opportunism and personal taste must play a major part in their choice.

According to Cocco (1987) the leaves of several palms may be used in the same way as *Geonoma baculifera*, including *Socratea*, *Bactris* and *Mauritia*. He also observed the use of many other plants for decoration, including the attractive flowers of *Ceiba pentandra*, and Fuentes (1980) reported body adornment with the flowers of *Amasonia arborea*, *Bauhinia glabra*, *Celosia argentea*, *Cydista aequinoctialis*, *Guarea guidonia*, *Hippeastrum* sp., *Passiflora longiracemosa*, *Spathiphyllum* sp. and *Tabebuia* spp. in his study area. He also mentions the young red leaves of *Licania kunthiana*, and an unidentified species of Annonaceae. Lizot (1984) cites the use of the attractive and aromatic flowers of *Randia* (*Tocoyena*) sp., and according to Finkers (1986) *Lantana trifolia* is grown in some villages specifically for the purpose of adornment.

Plants for construction

Temporary shelters

The most basic type of Yanomami shelter is the **naa nahi** (commonly known in Brazil as *tapiri*[41]), which is used in the forest by travelling groups or hunting parties (Fig. 33). This simple temporary structure plays a crucial

[41] From the Tupi language.

Figure 33. An abandoned temporary forest shelter. The *Phenakospermum* leaves used to cover it are already disintegrating.

Figure 34. Closed roundhouse at Palimiú on the Alto Rio Uraricoera.

role in the life of the Yanomami, who would traditionally have spent between one-third and one-half of the year on the move (see above). The **naa nahi** consists of a tilted planar grid of sticks, supported by tree trunks or slender poles lashed with vines or fibrous bark, and roughly roofed with large leaves (see Lizot, 1984 for an illustration). It is just large enough to shelter a small family and their cooking fire. Its temporary nature makes careful choice of the materials used to build it unnecessary. Thus, for example, the giant leaves of *Phenakospermum guyannense* and the smaller leaves of *Heliconia* spp. and *Calathea* spp. may be used for convenience as a thatching material, whereas for longer-term structures these species are unsuitable on account of their tendency to curl, split and disintegrate. Less favoured lashing materials may also be used for temporary shelters, such as the split stems of *Bauhinia guianensis* vines.

Structure and composition of the *yano*

The typical layout of a Yanomami round-house (**yano mat^ha** or **xapono**) has been described and illustrated in some detail by Chagnon (1968), Fuerst (1967) and Lizot (1984). Houses vary in size but are always round, with an opening in the centre. The size of this opening (**yano kahiki**) increases as the diameter of the house increases, and in the smallest 'closed' houses this is reduced to a small smoke-hole, sometimes capped by a type of thatched chimney (Fig. 34).

The materials and methods employed in the construction of the large round-house at **Watoriki** village (Fig. 35) were surveyed quantitatively in 1994. This **yano**, whose structure provides a good example of Yanomami architecture (although among the eastern upland Yanomami closed houses **yano komi** are more common), consists of a covered ring-shaped structure about 80 m in diameter, walled on the outside and open on the inside with a large open space (**yano a miamo**) in the middle (Plate 10b). It was built in a clearing (**yano a roxi**) just large enough to ensure that the tallest trees in the adjacent forest would not cause damage if they fell.

The outer wall at **Watoriki** is broached by four main doors, which are separated from the adjacent living areas by short walls. These principal openings[42] lead directly to the main trails from the village to the nearby streams, to the gardens and to the other Yanomami villages (Fig. 36). A number of other, smaller, doors (**wai yoka**) are used specifically by the families who live alongside them. The roof of the ring is made up of two parts. The main roof, which covers the living area, overlaps the inner (secondary) roof (Fig. 37), preventing rain from entering but allowing the smoke of the cooking fires to escape.

---

[42] Main doors (**pata yoka**) are classified according to where they lead: **hwama yoka** (guest doors) and **periyo yoka** (trail doors) open onto the principal paths where visitors and travellers enter and leave the village. There are also **rama yoka** (hunting doors) where hunting paths leave, **napë yoka** (strangers' doors) where paths lead to white settlements, **hutu yoka** (garden doors) where paths lead to the gardens, and water doors where paths lead to streams (**māu uka yo**).

Figure 35. The *yano* at *Watoriki*, with granite outcrop behind.

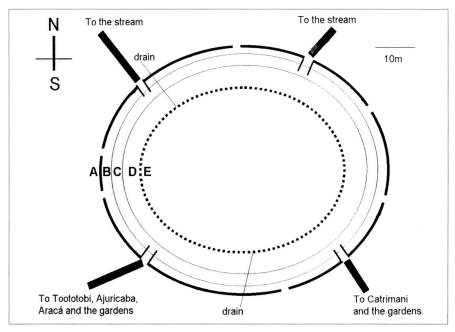

Figure 36. Aerial plan of the *Watoriki* round-house, showing the doors, principal trails and living areas. A = *yano a roxi* (clearing around *yano*), B = *yano a xikā* (feminine area); C = *nahi* (hearth, family space); D = *yano a hēhā* (masculine area); E = *yano a miamo* (central area).

Figure 37. Interior of the *yano* at *Watoriki*. Numbers correspond to the numbered components in Table 5.

The roofed area[43] is approximately 10 m in breadth, of which a little less than half is used as living space, occupied by clusters of hammocks around cooking fires and by racks and shelves on which food and a few belongings are stored. The inner half of the roofed area is kept clear and is used for communal and ceremonial activities and as a corridor. These zones correspond to the spaces between the concentric rings of posts shown in Fig. 36. The overall atmosphere is one of airiness and space, providing an extremely pleasant living environment.

During the roofing process a rough scaffolding, *yano iraki*, is erected. The roof is thatched with the fish-tail-shaped leaves of *Geonoma baculifera*[44], a small lower-understorey palm. These are folded in half and tied by the rachis, closely overlapping, to thin lengths of split *Socratea exorrhiza* trunks which are laid horizontally across the rafters and secured with *Heteropsis* roots[45].

[43] This includes the outer 'female' space behind the women's hammocks (*yano a xikā*), which is continuous with the hearth area occupied by the men's and children's hammocks (*nahi*). The portion between the hearths and the central opening is known as (*yano a hēhā*).

[44] According to the Makuxi, the length of time which these leaves will last depends upon the phase of the moon when they are cut (WM pers. obs.).

[45] These roots are used as the main lashing material throughout the house. However, the aerial roots of the epiphytic climber *Thoracocarpus bissectus* and the stems of the lianas *Callichlamys latifolia* and *Arrabidaea* sp. were also said to be suitable for this purpose.

Table 5
Inventory of the wooden components* of the *Watoriki* round-house, listed by tree species

Family	Species	Rafters		Tie-beams				Purlins		Posts				Total
		1	2	3	4	5	6	7	8	9	10	11	12	
Annonaceae	*Anaxagorea acuminata*	1												1
	Duguetia lepidota	4			6									10
	Fusaea longifolia	1	2	4	3	6							2	18
	Guatteria sp.	47	44	1		3	1	2						98
	Xylopia sp.	366	91	24	34	16	10	17	12					570
Apocynaceae	*Aspidosperma* sp.				1	1	1					1		4
Bignoniaceae	*Tabebuia capitata*											1		1
Burseraceae	*Protium fimbriatum*	1												1
Chrysobalanaceae	*Couepia caryophylloides*		6			1							23	30
	Licania aff. *heteromorpha*	3	5				1							9
	Licania kunthiana		3											3
	Licania cf. *polita*	2	11											13
Elaeocarpaceae	*Sloanea macrophylla*		1											1
Euphorbiaceae	*Croton matourensis*	1							1					2
	Maprounea guianensis		1											1
	Pogonophora schomburgkiana	1										2		3
Flacourtiaceae	*Casearia guianensis*		3			1		1	2				10	17
	Casearia javitensis		3											3
Lauraceae	*Aniba riparia*	2	5			5	4		3		8	4	25	56
	Licaria aurea		1											1
	Nectandra sp.	1												1
Lecythidaceae	*Eschweilera coriacea*												1	1
Leguminosae	*Centrolobium paraense*									17	15	23	6	61
	Martiodendron sp.		1									1		2
	Tachigali myrmecophila	58	30	7	2	10	10	11	4				4	136
	Zollernia paraensis										1			1
Meliaceae	*Trichilia* sp.				1								7	8
Monimiaceae	*Siparuna decipiens*				1								12	13
Moraceae	*Pourouma ovata*		2											2
	Pourouma tomentosa ssp. *persecta*	2	3											5
	Pseudolmedia laevis								1					1
Myristicaceae	*Iryanthera laevis / juruensis*	20	129			2		1	15				7	174
	Virola elongata	2	3											5
Myrtaceae	*Eugenia flavescens* vel. aff.			1	6		1			4	2	1	75	90
	Eugenia sp.												1	1
	Myrcia sp.	1		3	3	5				12	3	4	32	63
Palmae	*Bactris monticola*	1	1											2
	Socratea exorrhiza							9						9
Quiinaceae	*Quiina florida* vel aff.	1		2	3						1			7
Rubiaceae	*Duroia eriopila*	1					2			2			2	7
Sapotaceae	*Chrysophyllum argenteum*										1		11	12
	Manilkara huberi		1							12	22	43		78
	Pouteria caimito										1			1
	Pouteria cladantha	2	2	1		2				3	2		3	15
	Pouteria hispida												1	1
	Pouteria venosa ssp. *amazonica*												7	7
Violaceae	*Amphirrhox surinamensis*												55	55
	Rinorea lindeniana		1										23	24
Unidentified				1			2							3
		518	349	44	58	54	32	32	47	50	56	79	308	1627

* Column numbers correspond to the numbered components in Fig. 37.

Figure 38. Interior of the *yano* at **Watoriki**.

Thatching one square metre of roof requires approximately 160 *Geonoma* leaves, and an estimated 500,000 leaves were used for the whole building. To prevent the small leaves from being displaced or damaged in strong winds, fronds of the larger palms *Maximiliana maripa* and *Jessenia bataua* are attached above the *Geonoma* thatch.

A quantitative list of the 52 species used for timber, thatch and lashing in the construction of the **yano**, and the components for which they were employed, is given in Table 5. Most of the principal wooden components of the house had been stripped, but the bark had been left on the rafters between where they crossed the outside tie-beams and their outer ends, presumably to help protect them from the weather.

In the choice of wood for rafters, length (approx. 9 m for the outer roof), straightness, strength and lightness are all significant factors. Trees of the Myristicaceae, Annonaceae and (in some cases) Leguminosae are particularly well suited for this purpose, partly on account of their typical structures (architecture)[46]. The Annonaceae and Myristicaceae accounted for 81% of the 867 rafters (64% and 17% respectively). Young individuals of *Tachigali myrmecophila* (Leguminosae) had also been used in considerable numbers.

---

[46] Of these, Hallé *et al.* (1978) assign *Iryanthera* and *Virola* (Myristicaceae) to the Massart model of tree architecture, *Xylopia* (Annonaceae) to the Roux model and *Tachigali* and *Sclerolobium* (Leguminosae) to the Petit model, all of which are defined as having 'monopodial orthotropic trunk axes with plagiotropic branches' (straight unbranched trunks with horizontal branches).

Figure 39. Detail of the thatch at **Watoriki**. Note the bamboo quiver suspended from the roof.

The preferred tree for rafters is a *Xylopia* species, which makes up 53% of the total number and 71% of those of the outer roof. These figures differ because the outer roof, which is the most important since it covers the living area, was built first. When it came to the building of the inner roof, *Xylopia* trees were scarce in the vicinity of the village and a greater proportion of other species (mainly *Iryanthera* spp.) were used.

The requirements for the tie-beams, which must also be slender and strong but which are generally considerably shorter than the rafters, are met by most of the species mentioned above. Again, *Xylopia* and *Tachigali* are strongly represented, but the lesser need for length appears to have allowed some of the harder species (more commonly used for posts) to be used, whereas the greater need for strength has perhaps precluded the use of certain rafter species such as *Iryanthera* and *Virola*.

The roof is supported by concentric rings of posts (**yano nahiki**), of which the innermost three are the most important structurally. The posts must be very strong, and, as they are partially buried, resistant to rotting. Close-grained hardwoods are used for these posts, and of the 185 in the inner rings at **Watoriki** village, 97% were made from the wood of the four families Sapotaceae (45.5%), Leguminosae (31%), Myrtaceae (14%) and Lauraceae (6.5%). The most important species were *Manilkara huberi*, which accounted for 42% of the posts, and *Centrolobium paraense* (30%). The outermost ring was represented by a greater variety of species (21), probably because it is made up of a far greater number of posts (308), individual strength thus being of less importance.

The outer wall of the **yano** is built either of split sections of the trunk of *Socratea exorrhiza* (which is particularly easy to split), or of thatch of the type used for the roof (Fig. 40). *Socratea* is also used for the short walls beside the four main entrances. The composition of the wall and of other structural components varies around the perimeter, demonstrating the individuality of the family groups living beside it. The family which is to live in a particular section of the round-house is largely responsible for its construction, so the species used in any area will depend to some degree upon personal preference. In the case of the outer wall this may reflect whether the inhabitants lived through periods of inter-village skirmishes, when the strength of the wall (or in some cases of an outer palisade) as a defence was of greater importance than it is now. According to Smole (1976), these outer palisades, which are more common in the Parima highlands, would be maintained only when raids were expected. Fuerst (1967) described walls at Toototobi which were composed of an inner layer of *Socratea* planks and an outer layer of thatch, thus serving both as an effective wind-break and as a substitute for a palisade.

Fuentes (1980), in his general studies of the plants used by the Yanomami in Venezuela, listed the Yanomami names of 11 trees which were commonly used for house construction. Not all of these were identified, but they included the genera *Centrolobium*, *Duguetia*, *Eschweilera*, *Tabebuia* and *Tachigali*, all of which were recorded in the present study, as well as one member of the Burseraceae. Four plants were recorded as used for lashing, including *Heteropsis* and *Cydista* (Bignoniaceae), the second of which was not recorded at **Watoriki** but corresponds to the other bignoniaceous lianas collected there. In his travels among the upland Yanomami, Prance (pers. comm.) recorded a much greater use of the wood of *Eschweilera* spp. for posts than was found at **Watoriki**, where it was virtually absent. Lizot (1984) cites the use of eight preferred trees from his study area in Venezuela, and the occasional use of a further eight species. Again, the majority of these were not identified but they included *Duguetia*, *Eschweilera*, *Tachigali*, *Guarea*, *Geissospermum* and *Pera*, the last three of which were not used at **Watoriki**.

Changes in Yanomami construction

The **yano** described at **Watoriki** is largely typical (structurally) in the context of traditional Yanomami dwelling construction, but the choice of species used to build it has been adapted to meet the demands of small changes in the group's lifestyle. Certain building techniques have likewise been modified for the same reasons. In fact, differences in the species employed by the Yanomami for house construction are apparent right across their territory. To some extent these are the result of regional variations in the flora, but they are also influenced by other factors. The advent of steel tools, for example, will inevitably have affected the Yanomami's choice of building materials, allowing them to use harder woods than would previously have been worth the effort of felling.

Figure 40. One of the main doors to the **yano** at **Watoriki**. Note the thatched wall on the left, and the wooden wall (*Socratea exorrhiza*) on the right.

Traditionally the Yanomami tend to abandon their houses after a few years and to move on to a new area, partly as a result of the diminished populations of game animals in the vicinity, and in some cases the limited amount of land available at close range for swidden cultivation. The people of *Watoriki* lived for approximately five years in their last *yano* before moving to the present site, which was an average time for a group to remain in one place, these moves generally occurring after a minimum of 2–3 years and a maximum of 5–7 years. However, they have now made a conscious decision to remain at the new site for a longer period, largely on account of its current proximity to a FUNAI post and CCPY medical centre and the advantages which these are perceived to offer (medicines, trade goods, radio, air transport to the city). The community has therefore needed to give greater consideration to the properties of the materials used in the house's construction, such as the rot-resistance of the support posts, which will have to survive partial burial for longer than would normally have been necessary. It is therefore no coincidence that both *Centrolobium paraense* and *Manilkara huberi*, the preferred species today, are widely recognized in the timber trade as extremely resistant to rot (Rizzini, 1971).

To compensate for the increased pressure which this sedentariness tends to place on their natural resources, they have also begun to develop a system of secondary houses[47] which, although lying relatively close to the main house at Demini, open up a larger area of forest for intermittent exploitation. This new mobility system has virtually replaced the traditional system which revolved around hunting camps and trekking.

In discussions of the construction of the *yano* and the necessity of maintaining it over a relatively long period, the major preoccupation among the Yanomami interviewed was with the thatch and the availability of *Geonoma* leaves for its replenishment (Fig. 42). The gathering of the half million or so leaves for its initial thatching was evidently a mammoth task, and it seems that *G. baculifera* can take a long time to regenerate. This is supported by the observations of Balée (1994) in his study of the ethnobotany of the Ka'apor of the Eastern Amazon, who cites this species as one of only two which are directly threatened with 'micro-local extinction [extirpation] by traditional Ka'apor forest utilization'. In some of the longer-established Ka'apor settlements the *Geonoma* palms in the surrounding forests have been so depleted that they have had to resort to using other palm genera for thatching.

The thatching of the roof of the *yano* at *Watoriki* provides a clear example of the dynamic nature of Yanomami construction techniques. The houses used by this group formerly consisted only of an outwards sloping roof and a back wall, as is still the case among most other Yanomami communities in the region (as in Fig. 41). The inner roof, which gives the advantage of shade throughout the

[47] By the end of 1997 the *Watoriki* had built two of these secondary houses, one near the Ananaliú river (as a base for big fishing) and one approximately two hours walk to the northeast of the main village, as a base for hunting (Fig. 41).

Figure 41. One of the 'secondary' villages built for occasional use by the people of **Watoriki** in 1995.

day[48], was a relatively recent introduction borrowed from the eastern Yanomami village of **Kapirota u** situated on the Jutaí, a tributary of the Demini river. The thatching technique described above was said to have been learned from workers employed at an SPI post established on the upper Demini in the early 1940s when a border commission (CBDL) team came to the region. These workers were generally from other Indian groups such as the Tukano, and helped to build houses and camps using their own traditional thatching techniques.

At that time the upland Yanomami were using the leaves of *G. deversa* for their thatch, supported by the aerial roots of *Heteropsis* – a technique also in current use by the Maiongong in the uplands to the north – and it was only when the **Watoriki** people moved down into the lowlands, where *G. baculifera* (a better thatching species) is found in greater abundance, that they adopted the method used today. This is the same technique described by Fuerst (1967) from a *yano* on the Toototobi river.

Although tradition obviously plays an important part in the choice of materials for house construction, opportunism is inevitably an important factor as well. There is a limit to the distance which any piece of wood is worth carrying, however pressing the needs for durability, and the composition of the forest in the immediate vicinity of the chosen site will evidently influence the composition of the house (or conversely the location of the house may be

[48] In buildings without this inner roof, shade may be provided by banana leaves or woven palm-leaf matting, hung from the roof peak.

Figure 42. Repaired thatch at *Watoriki*.

affected by the composition of the forest). In the four quantitative Amazonian ethnobotanical studies discussed by Prance *et al.* (1987), the percentage of species in one hectare which were regarded as suitable for construction varied from 30.3 (Tembé) to 2.9 (Panare). The degree to which the house will be built of 'ideal' species will be determined by the abundance of those species in the area, by the necessity for an 'ideal' house, and by the manpower and technology available.

Table 6
Some tree species formerly used for construction by the people of *Watoriki* when they lived in the uplands

Species	Yanomami name	Family
Aniba riparia	*tʰua mamo hi*	Lauraceae
Dialium guianense	*paroko xihi*	Leguminosae
Euterpe precatoria	*maima si*	Palmae
Licaria aurea	*hōkōma hi*	Lauraceae
Maprounea guianensis	*yipi hi*	Euphorbiaceae
Martiodendron sp.	*rapa hi*	Leguminosae
Ocotea sp.	*iroma sihi*	Lauraceae
Pouteria sp.	*poxe mamokasi hi**	Sapotaceae
Pouteria venosa	*māiko nahi*	Sapotaceae
Quiina florida vel aff.	*naxuruma ahi*	Quiinaceae
Sclerolobium sp.	*paya hi*	Leguminosae

* The preferred construction species

Yanomami construction in the Amazonian context

One can observe considerable similarities between the Yanomami and other Amazonian peoples in the choice of plant species for construction. At the Waimiri Atroari village of Maré described by Milliken *et al.* (1992), for example, the rafters were made from the wood of various species of Annonaceae (including *Xylopia*), Myristicaceae (including *Iryanthera* spp.), and *Tachigali myrmecophila*, as they were at **Watoriki**. The main support posts were all made from the very hard and resistant wood of *Minquartia guianensis*[49], which is much used for this purpose in Amazonia but apparently absent from the forest around **Watoriki**. The outer ring of posts at Maré, as at **Watoriki**, was built from a broad range of genera including *Eschweilera*, *Licania*, *Pouteria*, *Quiina* and various Lauraceae and Meliaceae. As with the Yanomami house, the main lashing material used at Maré was *Heteropsis*.

Heteropsis roots (*cipó titica* in the Brazilian vernacular) are, by virtue of their strength, flexibility and widespread abundance, probably the most commonly used lashing material in Amazonia. However, it is also not just the Yanomami who employ bignoniaceous and cyclanthaceous vines in house construction. According to Boom (1987), the Chácobo of Bolivia use *Anemopaegma*, *Arrabidaea*, *Cydista* and *Melloa* species (Bignoniaceae), as well as *Thoracocarpus bissectus* (Cyclanthaceae). Likewise various Bignoniaceae, including *Anemopaegma*, *Distictella*, *Martinella* and *Mussatia* spp., have been used for rope in Guyana in the absence of better species (Fanshawe, 1948), and Glenboski (1983) and Barfod and Kvist (1996), cite the use of certain Bignoniaceae and Cyclanthaceae by the Tikuna of Colombia and the Cayapa of Ecuador.

The two most important plant families employed in construction at **Watoriki** (in terms of numbers of species used for the wooden components) were the Annonaceae and the Sapotaceae, which were also the most important families in the lists of Peruvian construction materials presented by Parodi (1988) and Pinedo-Vasquez *et al.* (1990) (see Table 7). As with the Yanomami, the Sapotaceae were used principally for support posts, and the Annonaceae for rafters and beams at these locations. In another Peruvian study (Phillips and Gentry, 1993), an analysis was made of the numbers of species employed for each of the principal components of houses. One hundred and twenty-two tree species were cited as suitable for beams and rafters etc., as compared to only 44 for house posts. Balée (1994) likewise reported that 48 species are used by the Ka'apor for roof parts but only 10 for posts. Similar proportions were found at **Watoriki**, where 38 species were used for the roof components and 14 for the three principal load-bearing rings of posts.

The extent to which a tree trunk will resist rotting and termite attack depends on both chemical and physical factors. The wood (secondary xylem) of certain species of *Eschweilera*, for example, contains silica (Ter Welle, 1976), and it seems that this provides some degree of resistance to termites. Many Chrysobalanaceae (e.g. *Licania*) also contain silica, and Balée (1994) suggests

---

[49] This species is so prized by the Tembé of Pará that there is a taboo on its burning, the breaking of which is said to result in deaths in the village (Balée, 1987).

that this is the reason for these families frequently being employed as rafters and beams by the Ka'apor. Similarly, the phenolic compounds found in various members of the Lauraceae and Sapotaceae (see Schultes and Raffauf, 1990) may be responsible for their resistance to fungal rot.

Table 7
Comparison of the principal plant families employed in house construction at *Watoriki* (numbers of species used) with those from three studies in Peru

Watoriki	Pinedo-Vasquez *et al.* (1990)	Parodi (1988)	Phillips & Gentry (1993)*
Sapotaceae (6)	Annonaceae (12)	Annonaceae (14)	Lauraceae
Annonaceae (5)	Palmae (5)	Palmae (12)	Myristicaceae
Palmae (5)	Sapotaceae (4)	Sapotaceae (6)	Annonaceae
Chrysobalanaceae (4)	Humiriaceae (2)	Leguminosae (6)	Leguminosae (Caesalpinoideae)
Leguminosae (4)	Melastomataceae (2)	Lauraceae (5)	Meliaceae
Euphorbiaceae (3)	Moraceae (2)	Moraceae (5)	Leguminosae (Papilionoideae)
Moraceae (3)	Myrtaceae (2)	Guttiferae (3)	Burseraceae
Myristicaceae (3)	Rubiaceae (2)	Cyclanthaceae (3)	Guttiferae

* Listed by descending 'family use value'

A comparison of the tree species used at ***Watoriki*** with those chosen by the Chácobo of Bolivia (Boom, 1987) again shows considerable similarity at the generic level. There, the genera employed for posts included *Aniba*, *Eschweilera*, *Dialium*, *Manilkara*, *Ocotea* and *Pouteria*, and those used for rafters etc. included *Anaxagorea*, *Aspidosperma*, *Duguetia*, *Guatteria* and *Xylopia*. These ethnobotanical correlations, which could probably be found between semi-nomadic peoples right across the forested regions of the Amazon, are more obvious than those for other plant uses (e.g. medicines). This is no coincidence, since anybody who is involved in the manual felling of patches of forest for swidden clearings (and subsequently observes the rates at which they rot in the gardens) will inevitably be well aware of the physical and structural properties of trees and vines, and it is natural that optimization will lead to the use of the same or related species.

Plants for tools, implements and miscellaneous uses

At ***Watoriki*** village, at the time of this study, a variety of industrially manufactured items were in use alongside traditional Yanomami technology. However, only in a very few cases (e.g. cutting implements and cooking pots[50]) had the traditional Yanomami items been almost entirely supplanted by their manufactured counterparts.

[50] See Lizot (1974) for a detailed description of traditional Yanomami pottery techniques.

Plate 1

Woman carrying cassava bread

Plate 2

A. Young boy practising archery.

B. Boy fixing an arrow-head to his arrow.

Plate 3

A. Women on a fishing expedition, with baskets of *Clibadium* leaves.

B. Fishing with a ***xotehe*** basket.

Plate 4

A. Seiving cassava flour with a square seive.

B. Mother carrying her child in a bark-fibre sling.

Plate 5

A. Boy playing in the centre of the **yano** with macaw feathers on a string.

B. Collecting an epiphytic medicinal plant (*Philodendron* sp.).

Plate 6

Cutting up a freshly killed deer on banana leaves.

Plate 7

Boys of *Watoriki* looking down on their *yano* from the weathered granite outcrop above.

Plate 8

A. Roberto relaxing in his hammock. Circular pieces of manioc bread are piled on the storage rack above him, together with his other possessions.

B. Woman scooping stunned fish from a stream with a loose-woven basket, her baby slung across her chest.

Plate 9

A. Women of Balawaú in the forest, one carrying a medicinal plant. Traditionally, the knowledge of medicinal plants is largely the preserve of the women.

B. A hunter carries a macaw home on his back, with a *Heliconia* leaf tied between the bird and his skin.

Plate 10

A. *Watoriki* from above, showing a recently created garden in the background.

B. The open space inside the yano at *Watoriki*, showing the overlap between the outer and inner roofs (where the smoke of the cooking fires escapes), and the large granite outcrop behind the village.

Plate 11

A. Children in the *yano* at *Watoriki.*

B. Yanomami (Xiriana) woman with her baby on the bank of the Rio Uraricoera. The unfinished canoe in the background is being soaked after being partially burned – part of the boat-building process employed by the neighbouring Ye'kuana.

Plate 12

A. The opened fruit of *Bixa orellana*, showing the bright red aril around the seeds, which is used as a dye by the Yanomami.

B. Antonio with calabashes of *açai* drink at a **reahu**.

Fibres

The strong bast fibres from the inner bark of certain trees are used for a variety of purposes. Broad strips of fibre from the trunks of *Brosimum utile*, *Sterculia pruriens* and *Cochlospermum orinocense*[51] are employed as slings for carrying babies (Plate 4b). *Guatteria* spp. may also be used for this purpose (Fuentes, 1980). When taken from small (young) trees, the tubular bark-cloth is removed whole from the trunk like peeling the stocking from a leg, in the same way that the Waorani of Ecuador are described as removing *Ficus* bark to make their baby-slings (Davis and Yost, 1983). The slings, which are sewn into a loop with silk-grass twine, tend to be dyed red with *Bixa orellana*. *Sterculia pruriens* is also the preferred species for these slings in Guyana, and its fibres are reputed to be extremely strong and versatile (Fanshawe, 1948).

Strips of the inner bark from *Anaxagorea acuminata*, *Croton matourensis*, *Eschweilera coriacea* and from young individuals of *Couratari guianensis* are used to make simple disposable hammocks (**okotoma nasiki**), useful for travellers (Fig. 43). Lizot (1984) also reports the use of the fibres from the aerial roots of an unidentified palm for this purpose, and Fuentes (1980) cites the use of *Duguetia* bark likewise. The bark of *Stryphnodendron pulcherrimum*, *Croton matourensis* or *Apeiba?* sp. is stripped in a single sheet (**naposi**) from the tree and used as a mat onto which the large quantities of cassava are grated during a **reahu** feast. Bast fibres from the aforementioned species and others, including *Elizabetha leiogyne*, *Fusaea longifolia*, *Guatteria* spp. and *Theobroma bicolor*, are also used to make carrying straps for baskets, game, etc. These tend to be stripped from the nearest trees with fibrous bark whenever a piece of cord is needed. Loops of bast fibre or the stems/aerial roots of flexible vines are also used as disposable foot-loops (*peconhas* in Portuguese) for tree climbing (Plate 5b).

The aerial roots of certain araceous epiphytes (*Philodendron* spp.) may also be used for general cordage, as may the stems of certain tough lianas such as *Bauhinia guianensis* and *Machaerium macrophyllum*. The bark of *Cecropia* spp. is used to make ropes for hanging hammocks etc., as are the fibres from silk grass leaves (*Ananas* sp.). Fibres from the bark of *Ficus*, *Couratari* and *Duguetia* spp. may also be made into ropes (Fuentes, 1980). Many of the species or genera which serve as sources of fibre for the Yanomami are used in the same way by other Amazonian communities (see Oliveira *et al.*, 1991). Balée (1994), for example, cites the use of 40 fibre-bearing species among the Ka'apor, including *Anaxagorea*, *Apeiba*, *Cochlospermum*, *Couratari*, *Duguetia*, *Eschweilera*, *Guatteria* and *Sterculia* spp.

Hammocks for long-term use are made from cotton twine, woven in a loose mesh between two upright posts. The cotton is spun on a simple spindle, which may be made from the wood of *Jessenia bataua* or from the central nerve of the frond of *Maximiliana maripa*, and a piece of calabash (*Crescentia cujete*).

---

[51] The women of **Watoriki** professed not to know the name of this tree, since it had not been present in the highland regions from whence the group came. They were, however, aware of the useful properties of its bast fibre.

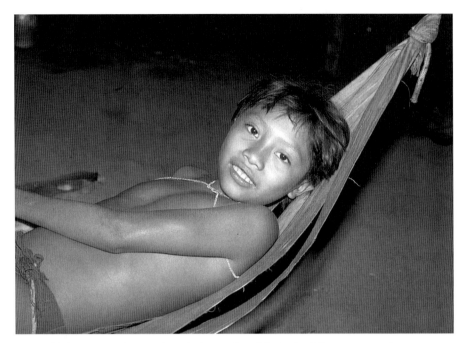

Figure 43. Resting in a hammock made from strips of the bark of *Croton matourensis*.

The cotton hammock is usually dyed red with *Bixa orellana* fruits. Hammocks may also be made from the split stems of *Heteropsis flexuosa*, bound tightly at each end with *Cecropia* fibre (Fig. 45). These hammocks, which are used by the western Yanomami at Balawaú, are more common among the Venezuelan Yanomami (see Lizot, 1984, for an illustration).

Basketry

Baskets are woven from the split (stripped) aerial roots of *Heteropsis flexuosa*, or, if this is in short supply, from the roots of *Philodendron* cf. *divaricatum*. These generally take one of four principal forms (illustrated in detail by Fuentes, 1980). The first (*wii*) is a large high-sided carrying basket, tightly woven with a rounded bottom, which fills the role of the open-backed *jamaxim* (usually made from *Ischnosiphon* spp.) favoured by the Carib tribes of the region. These are used by the women for carrying food, firewood etc. (Fig. 46), and may be dyed red with *Bixa* and decorated with simple black designs (spots and wavy lines).

The second basket type (*xotehe*) is a low-sided tightly-woven (*pesirimahe*) bowl, used as a receptacle for food. In these, the *Heteropsis* strips are sometimes interwoven with black fungal rhizomorphs (*uxiuxi kiki*) or with the dark stems of a wild grass called *hōrōma siki* for decoration, and some of the strips may be dyed purple with *Picramnia spruceana* leaves. *Heteropsis* roots are also used for basketry by the Waorani of Ecuador (Davis and Yost, 1983), and by the Maku. The third

74

Figure 44. Felicia spinning cotton, sitting in a traditional cotton hammock. The chequered hammock behind is a manufactured Brazilian one. Note the hexagonal-weave basket above, made from *Ischnosiphon arouma*.

Figure 45. Yanomami (Xiriana) woman from Palimiú, holding a *Heteropsis*-root hammock. Her intricate apron is woven from glass beads.

type of basket (also called *xotehe*) is much the same at the second, but is loosely woven (*wararimahe*) and generally undecorated. These can be used for food (lined with leaves) but also serve as sieves for scooping stunned fish out of poisoned streams (see Plate 3b) or for hunting small fresh-water shrimps, etc. When old they are used for gathering rubbish and throwing it out of the *yano*.

The fourth type (*sakosi*) is a simple cylindrical loosely-woven basket in a hexagonal weave. These baskets, which may equally well be made with split *Ischnosiphon* or *Heteropsis*, are produced in varying sizes and are used for storing food (lined with leaves), suspending small objects from the rafters and for transporting pet birds, etc. (Fig. 44). A small variation of this basket (*raekatamisi*), made with *Heteropsis*, is used as a sieve for cassava flour. Bast fibres or the split petioles of *Phenakospermum guyannense* and *Jessenia bataua* may also be used for weaving *sakosi* baskets (Fuentes, 1980; Lizot, 1984), as may the split petioles of young *Mauritia flexuosa* leaves (Cocco, 1987).

Figure 46. Carrying a *wii* basket.

Figure 47. Weaving a *tipiti*.

Whereas many Amazonian tribes use the split stems of *Ischnosiphon* spp. for most or all of their basketry, the Yanomami at **Watoriki** village use *I. arouma*[52] principally for weaving square loose-mesh sieves for sifting grated cassava (Plate 4a), and for tubular telescopic baskets (**koyotomasi**, known colloquially as *tipiti* in Brazil) for squeezing the poisonous juices from grated cassava prior to sieving (Fig. 13). The use of both the *tipiti* and the square sieve have been borrowed by the Yanomami, directly or indirectly, from neighbouring Carib tribes. In the case of the **Watoriki** the *tipiti* has been in use for a considerable period, but the sieve was introduced more recently as a consequence of the development of friendly relationships with the Yanomami of Paapiú (on the Couto de Magalhães river), who had contact with the Yanomami (Xiriana) of Ericó (on the Uraricaá) and the middle Mucajaí, who in turn had contact with Carib tribes. In the past, when bitter cassava varieties were processed, the juices were squeezed out between the hands (see Colchester, 1984) or between the knees in a **xotehe**-like basket (**ikatoma** – still in use), and then pressed again in a kind of bag made of cross-woven strips of *Anaxagorea* fibres (no longer used).

[52] Although only this species was collected at **Watoriki**, *Ischnosiphon obliquus* (Rudge) Koern. has been recorded elsewhere under the same name and for the same uses (BA).

Figure 48. Food cooking in the embers of the fire, wrapped in *Calathea*-leaf packets.

There are in fact several other types of baskets which are occasionally made. One of these (***xarapi***), square-woven with strips of the fibrous inner bark of *Couratari* sp., is described and illustrated by Fuentes (1980). Temporary backpacks, used for carrying game, fruits etc. home from the forest, are rapidly woven from two pinnate palm leaves, generally of *Euterpe precatoria* or *Jessenia bataua*. These packs (***paxoahi***) are carried by means of a head-strap made from bast fibre and tend to be used once only and discarded. *Geonoma* palm leaves are often used for wrapping miscellaneous items, and the leaves of *Heliconia* and *Calathea* (and doubtless other Marantaceae) are used to wrap small items collected in the forest, or food for cooking in the fire[53] (Fig. 48). They may also serve as linings for the otherwise rather porous palm-frond packs. Larger leaves (e.g. *Phenakospermum* and *Musa*) are placed on the ground as mats for cutting up game (Plate 6), or for dividing food at the end of ***reahu*** feasts. According to Fuentes (1980), when some of these leaves (including a much-used *Calathea* species), are used for cooking they impart a pleasant aroma to the food. Many aromatic Zingiberaceae are deliberately used in this way by the Yali highlanders of West Papua for cooking sweet-potatoes and other vegetables in their ground-ovens (WM pers. obs.).

---

[53] These leaf packets are called ***haro***. Young *Mauritia* leaves are also sometimes used to wrap fishes for cooking.

Other containers

Bowls (calabashes), now made from the cultivated fruits of *Crescentia cujete*[54], are dyed with the red resin from the trunk of *Licania heteromorpha* or *Inga alba*. *Couepia caryophylloides* resin is also used as a dye. These resins are generally mixed with lamp-black, producing the same lacquer that is used by numerous tribes of the Amazon to decorate their baskets (see Milliken *et al.*, 1992). According to Cocco (1987) this practice was learned from the Maquiritare (Ye'kuana). *Licania heteromorpha* is also used to dye calabashes by the Ka'apor in the eastern Amazon (Balée, 1994) as well as in French Guiana and Guyana (Grenand and Prevost, 1994), as are various species of *Inga*.

The old people at **Watoriki** described the use of *Euterpe* spathes to make wallet-like covers in which men used to keep their smaller fletching feathers. Larger feathers (e.g. macaw tail feathers) were kept in a bamboo tube (see Zerries and Schuster, 1974). Prior to the introduction of the calabash a small trough-like container (**axeposi**), made from palm spathes which were folded and then sewn at each end, would also have been used for food, and Lizot (1984) describes similar practices among the western Yanomami. Other soft palm spathe containers, from a palm of the same name or from *Socratea exorrhiza*, were used to heat water and cook palm fruits etc., and the tough *Maximiliana maripa* spathes (**okorasisi āthe**) were used as containers for flour and pulp during the processing of cassava. The use of the spathes of large palms as *ad hoc* cooking pans is also mentioned by Finkers (1986). According to Valero (1969), boxes for feathers etc. were also made from split *Socratea exorrhiza* wood, sewn together with vines.

Clay pots were commonly used for cooking in the past, and it was said that the clay was smoothed with the skins of the fruits of *Picramnia spruceana*[55]. These pots have largely been replaced by aluminium pans (Fig. 49), but the larger ones were still in use during **reahu** feasts in the 1970s and the early 1980s, and in remote areas some are still traded between villages. Large bottle gourds, made from the cultivated fruits of *Lagenaria siceraria*, were originally used for collecting water (now replaced by aluminium pans), and smaller ones are still used to store white down feathers for body decoration, liquids (e.g. *Bixa* body paints) and the crushed bones of the dead. Those for the latter purpose (painted black and red) are made from small rounded varieties known as **pora axi**, which is also the specific name for *Posadaea sphaerocarpa*[56]. Split longitudinally into two, gourds of all sizes serve as bowls for preparing or drinking liquid foods. The emptied fruits (pyxidia) of Brazil nuts (*Bertholletia excelsa*) serve as mortars for crushing food etc., and the

---

[54] This species was introduced to the Yanomami by employees of SPI/FUNAI and missionaries (probably between the 1940s and the 1960s).

[55] Why these fruits should have been used for smoothing clay pots is not clear.

[56] This apparent nomenclatural confusion is deliberate, resulting from the strong Yanomami taboo on speaking about topics related to death, requiring that they refer euphemistically to another type of gourd.

Figure 49. Justino with a drink made from *Oenocarpus bacaba* fruits, celebrating a visit from another village. Aluminium containers have replaced traditional earthenware pots. The visitors in the background can be identified by the white down glued into their hair.

hollow stems of large bamboos (*Guadua* spp.) are sometimes used as temporary water containers.

The trunks of *Ceiba pentandra, Cedrelinga catenaeformis, Licaria aurea* and *Rhodostemonodaphne grandis* are hollowed out to make large canoe-like containers for the preparation and storage of plantain soup (**koraha u**) for the **reahu** feasts, and smaller versions of these hollowed trunks are used for pounding the bones of the dead. Cocco (1987) mentions *Licania* wood as being suitable for these smaller 'mortars', and *Rinorea* and *Pouteria* for the crushing pestles. Troughs are also prepared from the flexible bark of *Croton* spp., folded and sewn at each end, and according to Finkers (1986) the swollen trunk of *Iriartea deltoidea*, a palm which was not observed by the authors but which seems to occur over only a small part of the Yanomami territory (Henderson *et al.*, 1996), is hollowed at the widest part and used as a trough.

River transport

The people of **Watoriki** and indeed the Yanomami in general are not a river people, and, generally speaking, do not use dug-out canoes except where they have been in long-term contact with riverine Carib peoples (see Plate 11b). It was said at **Watoriki**, nevertheless, that the wood of *Clarisia racemosa* is suitable for canoe building. This species is also used for dug-out canoes by the Waimiri

Figure 50. Gourds and calabashes suspended from the beams of the **yano**.

Atroari (Milliken *et al.*, 1992), and by the Wai-Wai and the Ye'kuana[57] (WM pers. obs.). The Ye'kuana claim that this wood is highly resistant, lasting up to 20 years, but that it has the disadvantage of sinking if flooded. According to Fuentes (1980), the bark of *Tabebuia guayacan* is used to make temporary canoes (see also the photograph in Zerries and Schuster, 1974), but these are generally suitable only for a single downstream journey. One informant at **Watoriki** also mentioned the use of the bark of *Bagassa guianensis* for this purpose.

Lizot (1974) describes the production techniques for these bark canoes in some detail, and concludes that although the Yanomami traditionally possessed these techniques for making their *Croton*-bark 'troughs' for banana soup, it was only under the influence of their neighbours the Ye'kuana that they adapted them to canoe building. Fuentes (1980) also describes the building of rafts from light, buoyant tree trunks such as *Didymopanax morototoni*, and the construction of simple bridges from the rot-resistant trunks of *Eschweilera coriacea*, *Tabebuia guayacan* and *Psidium* sp., with bignoniaceous or leguminous lianas (*Cydista aequinoctialis*, *Tynanthus polyanthus* and *Bauhinia* sp.) as hand-rails. Cocco (1987) likewise cites the use of *Cordia tetrandra* wood for rafts and *Elizabetha* wood for bridges.

Cutting tools

One of the principal cutting tools used traditionally by the Yanomami is the incisor of the agouti (*Dasyprocta* sp.), set into the end of a small piece of wood (e.g. *Rinorea lindeniana*) with silk-grass fibres and *Moronobea* or *Symphonia* resin. These were still in use in some areas at the time of study, usually kept in pairs

[57] The Ye'kuana (also called Maiongong or Maquiritare) share part of their lands with the Yanomami, and are expert boatmen and canoe builders.

(*tʰomi naki*) attached by a piece of string and carried on the back tied to the quiver. The tooth is traditionally sharpened with a piece of the inner bark of *Exellodendron barbatum* (Chrysobalanaceae), which contains silica and is therefore abrasive. The silica content of the bark of a number of Chrysobalanaceae (including *Hirtella*, *Licania* and *Parinari*) is well known amongst the Indians of South America, many of whom add its gritty ashes to clay in order to strengthen their pots (Beck and Prance, 1991; Schultes and Raffauf, 1990). For cutting softer materials (e.g. flesh, hair etc.), the Yanomami traditionally employed the razor-sharp edges of split *Guadua* bamboo stems such as **rahaka wana** ('quiver bamboo') and **rahaka si hrakarima** ('soft skin bamboo'), replaced a long time ago by steel knives (**poo ihuru**).

The people of **Watoriki** explained that the stone axe-heads (**manamo koxi**) found in their land are the relics of former occupations by another tribe (the Pauxiana, an extinct Carib group[58]) before their own migration into the lowlands. Nevertheless, prior to the advent of steel they would probably have used small stone tools themselves. Instead of the large stone axes used by other tribes, these probably took the form of small stone hatchets of the type described by Steinvorth de Goetz (1969) and illustrated by Zerries and Schuster (1974). One such hatchet was collected on the upper Catrimani by BA in 1975, having been brought back by warriors from a raid on an isolated group on the upper Apiaú river, known as the **Moxihatetemapë** or **Yawaripë**. According to the older people at **Watoriki** the blades of these hatchets, which were used for cutting small trees, opening up bees' nests etc., could also have been made in the remote past with pieces of the carapace of the giant armadillo *Priodontes maximus*.

The first steel used by the Yanomami probably arrived, generally in the form of worn or broken pieces of machetes or axes, through inter-tribal trading and/or spoils of war with groups who were in direct contact with the white frontier since the 19th century. These precious fragments were set into pieces of wood using *Cecropia* fibres and the guttiferous resins described above (Fig. 9), and employed as all-round hatchet-like cutting implements (**haowatima**)[59], now replaced by machetes (**poo pata**). This transition from stone to steel and its complex and interesting implications – not least upon the way in which the Yanomami exploit their plant resources – is discussed in some detail by Colchester (1984), and also above under 'Yanomami swidden horticulture'. Most communities now have access to manufactured steel axe-heads (**poo koxi**). The wood of *Aspidosperma nitidum* was said to be particularly suitable for handles for these axes, partly because of its strength but also because the furrowed or fenestrated nature of its trunk facilitates the cutting of a suitably shaped piece without a saw. This species is also used for axe-handles by the Waimiri Atroari (Milliken *et al.*, 1992), and by *caboclos* in Mato Grosso (J.A. Ratter, pers. comm.).

---

[58] The last Pauxiana were seen on the Catrimani river in the 1920s.

[59] However, such tools are sometimes still used: we saw an old man at Toototobi who was still using one made from a worn machete blade for opening bees' nests, and claimed that it is better than a steel axe and easier to transport and climb with.

Figure 51. Men contesting with fighting-sticks at Xitei.

Other uses

The strong and heavy wood of the saplings of *Mouriri nervosa* or *Duguetia* spp. is used in the highlands to make the long **rape tihi** clubs which are employed in ritualized inter-village fights (Fig. 51). Short paddle-shaped digging-sticks (**sihema**), made from the wood of *Jessenia bataua*, are also used as clubs for more improvized fights, and according to Cocco (1987) these could also be made from *Rinorea* or *Licania* wood. Spears were carved in the past from pieces of thick bamboo (*Guadua* sp.) or from the wood of certain palms (*Oenocarpus bacaba*, *Bactris gasipaes*). Less specifically, pieces of strong wood, chosen as much for their immediate availability as for any other properties, are used to make impromptu tools such as the simple hooks (**yopena**) for breaking fruits off trees branches (from the ground), or the ingenious crossed-pole climbing device used to gather fruits from spiny-trunked peach palms (see Cocco, 1987).

Further miscellaneous technological applications of plants include the use of pieces of the stem of *Olyra latifolia* as flutes/whistles at ceremonial occasions, and of the spiny fruits of *Apeiba membranacea* (or the spiny seeds of *Caryocar* spp.) as combs. The latex from the bark of *Manilkara* spp. is sometimes used as glue, the dried inflorescences of *Euterpe precatoria* are employed as brooms (Fig. 52), and the rough bark of *Couma macrocarpa* as a grater for cassava. This bark has now been replaced at **Watoriki** village, by

Figure 52. Sweeping with a dried *Euterpe* palm inflorescence.

pieces of pierced tin or by the superb graters made by the Ye'kuana[60]. According to Fuentes (1980) the spiny aerial roots of *Socratea exorrhiza* (Fig. 53) were also used as graters for food in the past. *Olyra latifolia* is also used for whistles by the Bora of Peru (Denevan and Treacy, 1987), and *Apeiba* and *Socratea* are employed as combs and graters (respectively) by both the Ka'apor and Waimiri Atroari (Balée, 1994; Milliken *et al.*, 1992).

The leaves of *Pourouma bicolor* ssp. *digitata* are used as sandpaper for smoothing wooden objects, and the young fronds of *Euterpe* or, according to Cocco (1987), *Oenocarpus* palms, are used in the ceremonial displays/dances which take place at the **reahu** feasts. *Clusia* vines are used for traditional tug-of-war contests, and needles for piercing the lips of young girls are made from the wood of *Jessenia bataua* or *Maximiliana maripa*. Valero (1969) also describes the use of thorns from the aerial roots of *Socratea exorrhiza* for this purpose. The Tikuna of Colombia likewise use young palm fronds (of *Astrocaryum chambira*) at girls' puberty ceremonies (Glenboski, 1983), and the Ka'apor and Wayãpi also use *Pourouma* leaves as abrasives (Balée, 1994; Grenand, 1980).

[60] These graters are made from a block of *Cedrela* (Meliaceae) wood, dyed red with *Bixa orellana* and set with numerous small stone (or nowadays metal) chips, which are fixed with a layer of the sticky latex tapped from the trunk of *Couma macrocarpa* (Apocynaceae). In the past, these graters reached **Watoriki** via long-distance trade channels, but they can now be obtained from the Ye'kuana living in Boa Vista.

For ridding dogs of fleas, they are bathed in an extract of the beaten stem of *Lonchocarpus* cf. *chrysophyllus* (more commonly used as a fish poison), and to keep biting blackflies (Simulidae) at bay the leaves of *Jacaranda copaia* are burned to produce a repellent smoke. The insecticidal activity of the rotenone found in *Lonchocarpus* is well known and has long been exploited commercially (Krukoff and Smith, 1937; Le Cointe, 1936), and the burning of *Jacaranda copaia* as an insect repellent has been reported from the Tapajós region of Brazil (Branch and da Silva, 1983) and among the Wayãpi of French Guiana (Grenand *et al.*, 1987). Fuentes (1980) reports a number of additional miscellaneous plant uses, most of which have also been recorded at **Watoriki**. Thus, the fruits of certain palms (including *Mauritia flexuosa*) are used to make spinning tops for children, the stinging leaves of *Urera* spp. or the spiny branches of *Acacia* sp. are used in traditional 'battles' between boys and girls, and *Euterpe* spathes are used as toy canoes.

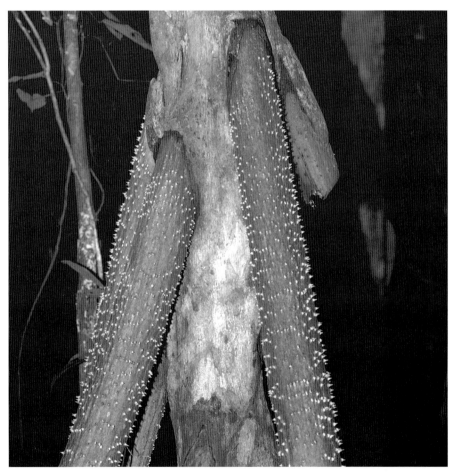

Figure 53. The spiny aerial roots of *Socratea exorrhiza*, which were used for grating cassava in the past.

Medicinal plants

The Yanomami pharmacopoeia

Many of the early anthropological studies made among the Yanomami (e.g. Chagnon, 1968) reported that they did not use medicinal plants (**hwëri mamo tima tʰëpë**) to any significant extent, but relied almost entirely upon shamanic healing (**xapirimou**). This belief has continued to surface in recent ethnobotanical works dealing with the Yanomami (e.g. Plotkin, 1993). However, it has now become clear that this is an erroneous assumption which probably grew out of the emphasis laid by earlier researchers on shamanic healing, and the consequent neglect or dismissal of the parallel practice of phytotherapy. Although the most important aspect of Yanomami healing practices is without doubt their shamanic medicine, at least in terms of cultural significance, there also exists a strong and well-developed practical tradition of the use of medicines prepared from plants and fungi and, to a lesser extent, insects[61].

An initial survey carried out at **Watoriki** village between 1993 and 1994 revealed the knowledge of at least 113 medicinal plant and fungus species among this group, which is in itself a very substantial number. Further research in the Xitei region in 1995 yielded 109 medicinal species, of which 80 had not been collected at **Watoriki**. Several of the 'new' species collected at Xitei (e.g. *Croton palanostigma, Swartzia schomburgkii, Tabernaemontana angulata*) are absent from the forests around **Watoriki** on account of the floristic differences between the two regions, and several of the species collected at both localities were used for different medicinal purposes. Additional data collected during a brief visit to Balawaú brought the overall total to 198 species. The substantial differences between the information collected at **Watoriki** and Xitei suggest that studies of other communities would yield considerable quantities of new information.

This pharmacopoeia is comparable with, or larger than, most of those recorded amongst other Amazonian indigenous peoples. Bennett (1992), for example, listed 245 medicinal species among the Shuar, Grenand *et al.* (1987) listed 180 among the Wayãpi, and Boom (1987) listed 174 among the Chácobo. Cavalcante and Frikel (1973) positively identified 171 among the Tiriyó and collected a further 157 specimens which were not classified, Balée (1994) listed 110 species among the Ka'apor, and Glenboski (1983) listed 84 among the Tikuna.

Although none of these figures claim to represent the total knowledge of the peoples concerned, they probably give some indication of the numbers of species employed. It appears that most of the surviving knowledge of medicinal plants has been collected at **Watoriki**, but the time spent in the field at Xitei and Balawaú during the present study was relatively short, and in

---

[61] Various insects with medicinal properties were collected during this study, including three species of bees, eight species of ants and two species of termites.

Table 8
Some medicinal plants used by the Yanomami[62]

Species	Family	Yanomami name	Medical use
Abuta rufescens	Menispermaceae	*werehe tʰotʰo*	Malaria
Anacardium giganteum	Anacardiaceae	*oru xihi*	Diarrhoea & stomach ache
Aristolochia disticha	Aristolochiaceae	*xuu tʰotʰo*	Diarrhoea & stomach ache
Aspidosperma nitidum	Apocynaceae	*hura sihi*	Malaria
Astrocaryum aculeatum	Palmae	*ëri si*	Fevers
Bauhinia guianensis	Leguminosae	*tüwakarama tʰotʰo*	Diarrhoea & dysentery
Caladium bicolor	Araceae	*xõwa a*	Wounds
Capsicum frutescens	Solanaceae	*prika aki*	Ophthalmia & respiratory infections
Cecropia aff. peltata	Moraceae	*tokori hanaki*	Carbuncles & abscesses
Clematis dioica	Ranunculaceae	*hemare mothoki*	Skin irritation
Clusia spp.	Guttiferae	*pori pori tʰotʰo*	Lesions
Costus guanaiensis	Zingiberaceae	*naxuruma aki*	Coughs
Croton palanostigma	Euphorbiaceae	*kotoporo sihi*	Fevers
Cymbopogon citratus	Gramineae	*makiyuma hanaki*	Pain
Cyperus articulatus	Cyperaceae	*haro kiki*	Fevers
Dieffenbachia bolivarana	Araceae	*xenoma a*	Ant stings
Drymonia coccinea	Gesneriaceae	*hurasi hanaki*	Fevers
Geophila repens	Rubiaceae	*mamo wai kiki*	Eye infections
Hippeastrum puniceum	Amaryllidaceae	*si waima a*	Stomach ache
Hymenaea parvifolia	Leguminosae	*arõ kohi*	Respiratory disorders
Inga acuminata	Leguminosae	*ria moxiririma hi*	Oral thrush
Machaerium quinata	Leguminosae	*rääsirima tʰotʰo*	Stomach ache & worms
Maranta arundinacea	Marantaceae	*hore kiki*	Wounds
Monstera adansonii	Araceae	*xãã a*	Abscesses
Nicotiana tabacum	Solanaceae	*pee nahe*	Botfly infestation
Peperomia magnoliifolia	Piperaceae	*në wãri hanaki*	Fevers
Peperomia rotundifolia	Piperaceae	*oru kiki wite*	Coughs
Philodendron solimoesensis	Araceae	*puu tʰotʰo*	Ant stings
Phlebodium decumanum	Polypodiaceae	*tokosi hanaki*	Coughs
Phytolacca rivinoides	Phytolaccaceae	*kripiari hi*	Infected chiggers
Picramnia spruceana	Simaroubaceae	*koeaxi hi*	Skin disorders & wounds
Piper arborea	Piperaceae	*kahu mahi*	Fevers
Pothomorphe peltata	Piperaceae	*mahekoma hanaki*	Malaria
Protium spruceanum	Burseraceae	*warapa kohi*	Respiratory disorders
Renealmia floribunda	Zingiberaceae	*nini kiki*	Pain
Siparuna guianensis	Monimiaceae	*mõe hi*	Dizziness
Spondias mombin	Anacardiaceae	*pirima ahi tʰotʰo*	Fevers
Swartzia schomburgkii	Leguminosae	*xotokoma hi*	Diarrhoea & intestinal pains
Tabernaemontana macrocalyx	Apocynaceae	*akiã hi*	Botfly infestations
Tabernaemontana sananho	Apocynaceae	*tʰoru hwãtemo hi*	Worms
Tanaecium nocturnum	Bignoniaceae	*puu tʰotʰo moki*	Skin irritation
Uncaria guianensis	Rubiaceae	*ërama tʰotʰo*	Diarrhoea & stomach ache
Urera baccifera	Urticaceae	*ira naki*	Body pains
Vismia angusta	Guttiferae	*yoasi hi*	Fungal skin disorders
Vismia guianensis	Guttiferae	*siiriama sihi*	Wounds & burns
Zanthoxylum pentandrum	Rutaceae	*naharã hi*	Toothache
Zanthoxylum rhoifolium	Rutaceae	*mamo wai hanaki*	Eye infections
Zingiber officinale	Zingiberaceae	*amatʰa kiki*	Coughs & toothache

[62] These species are selected on the basis of their having been recorded and published elsewhere with the same medicinal uses as were attributed to them by the Yanomami. The names of species whose medicinal applications appear to be particular to the Yanomami have been withheld in this and previous publications, in an effort to minimise the risk of violation of their intellectual property rights (see Milliken and Albert, 1996; 1997a).

neither of these locations could the species recorded be taken to represent the sum of the medicinal species known there. Without doubt the total number of medicinal species known to the Yanomami is very much greater than that which we have documented to date, and their pharmacopoeia may in time prove to be one of the most diverse recorded anywhere.

Figure 54. *Eleutherine bulbosa*, a cultivated medicinal species. Medicinal and magic plants tend to be planted at the base of tree stumps.

Most of the medicinal plants are wild forest species, but a few (16 spp.) are cultivated plants which are either grown specifically for their medicinal properties or are kept primarily for other important uses (food, fibre, etc.). The representation of botanical families among the plants collected at Xitei is very similar to that recorded at **Watoriki** (see Table 9), with particularly strong representation of the Leguminosae (*sens. lat.*) and the Piperaceae at the specific level. The small differences which are apparent in the representation of plant families between these two regions may be partly a consequence of the floristic variations discussed above. The majority of the families are represented by one or two species only, but there is a small group which consistently appears among the most strongly represented. In the overall list of Yanomami medicinal plants, 91 (46%) of the species are representatives of only 11 (14%) of the families.

The taxonomic composition of the Yanomami pharmacopoeia is fairly typical of those recorded in the northern Amazon. The Piperaceae, Leguminosae, Araceae, Rubiaceae, Moraceae, Annonaceae, Apocynaceae and Guttiferae are well represented in the Amazonian medicinal literature, all of them featuring amongst the 13 top families (by number of species used) in an overall survey of medicinal plants of French Guiana (Grenand *et al.*,1987). However, although this may be interpreted as a pointer to the most pharmacologically active families in the region, one must also bear in mind the relative sizes and diversities of those families, which will influence their representation. Nonetheless clear disparities between family size and 'family use value' have been demonstrated for medicinal plants in Peru by Phillips and Gentry (1993b).

Table 9
Principal medicinal plant families recorded among the Yanomami*

Overall (all study sites)		Xitei region		*Watoriki*	
Family	No. spp.	Family	No. spp.	Family	No. spp.
Leguminosae	13 (12)	Leguminosae	9 (9)	Leguminosae	9 (9)
Piperaceae	13 (3)	Piperaceae	8 (3)	Piperaceae	8 (2)
Araceae	12 (6)	Gesneriaceae	6 (3)	Araceae	6 (4)
Moraceae	9 (6)	Moraceae	5 (5)	Moraceae	6 (5)
Rubiaceae	9 (6)	Araceae	4 (4)	Rubiaceae	6 (6)
Guttiferae	7 (4)	Guttiferae	4 (4)	Zingiberaceae*	5 (3)
Annonaceae	6 (4)	Melastomataceae	4 (2)	Annonaceae	4 (4)
Gesneriaceae	6 (3)	Zingiberaceae*	4 (3)	Apocynaceae	4 (2)
Zingiberaceae*	6 (3)	Acanthaceae	3 (3)	Guttiferae	4 (3)
Apocynaceae	5 (2)	Burseraceae	3 (1)	Bignoniaceae	3 (3)
Menispermaceae	5 (3)	Myrtaceae	3 (2)	Monimiaceae	3 (2)
Burseraceae	4 (1)	Orchidaceae	3 (1)	Palmae	3 (3)
Melastomataceae	4 (2)	Rubiaceae	3 (3)		
Palmae	4 (4)	Solanaceae	3 (3)		

* Including Costaceae

The majority of the plant medicines recorded at **Watoriki** did not appear to be in use when the information was collected, although some may be resorted to when no alternative is available (e.g. on hunting and trekking expeditions). This discontinuation has primarily been due to the increasing availability of 'Western' allopathic medicines since the mid-1970s. The Yanomami probably established their initial faith in this introduced system as a response to its effectiveness against otherwise incurable epidemic diseases.

It was said at **Watoriki** that the knowledge of traditional plant medicine had originally been kept and practised by the older women. In general, particularly for treatment of fevers etc., this would have been practised *after* a shamanic healing session had been performed by men, as described by Valero (1984) among another Yanomami sub-group. There were, however, no survivors of the generation of women well versed in these traditional medicines remaining at **Watoriki**, the last old woman having died in 1984. Thus most of the knowledge of medicinal plants remaining within the group was that which a few of the older men had picked up from their mothers before the great epidemics of the 1970s took their toll on the older population.

With the exception of some of the diseases which have been brought to the region in the recent past by white people, the pharmacopoeia includes remedies for almost all of the medical disorders which the Yanomami would commonly encounter in their lives. The greatest diversity of plants are used for the treatment of fevers, stomach ache and intestinal pains, malaria, diarrhoea,

Figure 55. Carlita, one of the older women at Balawaú, explaining the use of a medicinal plant to a CCPY health worker. Note the *Geonoma*-leaf fibres she is wearing in her ears.

91

infectious epidemic diseases, coughs, eye infections, toothache, headache, snake bite, itching, intestinal worms, respiratory infections/congestion, body pains, oral thrush and ponerine ant stings[63].

Some of the boundaries between the illnesses and thus between the medicinal applications perceived by the Yanomami differ from those generally accepted in Western medicine, making it difficult to draw absolute parallels. For example, many plants are presented as treatments for the broad category **xawara a wai** (infectious epidemic diseases), under which are subsumed at least 18 specific diseases associated with the coming of white people, including measles, influenza, whooping-cough, etc. (see Albert and Gomez, 1997). At Xitei, **xawara a wai** is also known as **teosi a wai**, derived from *Deus*, the Portuguese word for God (the first white people in the region were evangelical missionaries). When a plant is recommended as a remedy for **xawara** this can generally be interpreted as meaning that the plant relieves one or more of the symptoms associated with this broad category, i.e. fevers, headaches, muscular pains, joint pains etc.

If one compares the applications to which the Yanomami put their medicinal plants with those of other groups, one again encounters considerable similarities (see Table 10). One cannot make a direct quantitative comparison

Table 10

Principal uses of medicinal plants among four South American peoples*

Yanomami	Spp	Chácobo	Spp	Miraña	Spp	Tiriyó	Spp
Fevers	60	Stomach ache	32	Skin disorders	23	Fevers	53
Stomach/intestinal pains	35						
Malaria (external use)	24	Skin infections	26	Fevers	19	Wounds & ulcers	21
Diarrhoea	23						
Epidemic diseases	15	Diarrhoea	25	Gastro-intestinal	19	Rheumatism	9
Malaria (internal use)	14						
Coughs	13	Rheumatism	25	Rheumatism	8	Headache	8
Eye infections	13						
Toothache	13	Toothache	10	Infections (bacterial)	10	Anaemia & weakness	7
Headache	11						
Snake bite	11	Hepatitis	10	Inflammation	8	Stomach ache	6
Itching	10						
Worms (intestinal)	10	Fevers	9	Pain (analgaesic)	8	Toothache	6
Congestion	9						
Respiratory infections	9	Vomiting	9	Wounds	8	Dizziness & vision	6
Body pains (localised)	8						
Thrush (oral)	8	Appendicitis	6	Bronchial disorder	5	Coughs	6
Ponerine ant stings	8						
Weakness/malaise	8	Headache	5	Snakebite	6	Convulsions	5
Leg pains (rheumatism?)	7						
Abscesses	7	Eye infections	4	Eye infections	3	Eye infections	4
Nausea	6						
Burns	5	Head colds	3	Liver	3	Hepatitis	4
Leishmaniasis	5						

* Data from Milliken and Albert (1997a), Boom (1987), La Rotta (1988) and Cavalcante and Frikel (1973)

[63] According to CCPY's 1997 medical data for Demini, Toototobi and Balawaú, the most commonly reported medical problems were flu (51% of cases), malaria (7%), complications of flu (6%), wounds/trauma (6%), conjunctivitis (5%), diarrhoea (5%), skin diseases (5%), and unknown fevers (4%).

between them, since the differences in the ways in which these groups (and the ethnobotanists who worked among them) perceive disease affects the way that the medicines are classified, but nevertheless there is a clear preponderance of certain problems such as fevers, stomach and intestinal disorders, bacterial and fungal infections of skin, wounds and eyes, respiratory disorders, toothache, etc.

Application and preparation techniques

Most of the plant medicines used by the Yanomami are prepared from the leaves or bark, although in some cases roots, sap (latex), fruits, flowers, stems and other parts are employed. Trees (70 spp.) and terrestrial herbs (46 spp.) are the most common sources of medicines, but shrubs (26 spp.), lianas (24 spp.) and epiphytes (22 spp.) are also important. Most plants are prepared and used singly, which is a common feature of indigenous Amazonian medicine. Of the 124 medicinal preparations recorded among the Tikuna by Glenboski (1983), for example, only 10 involved more than one species, and similarly the Wayãpi use most of their medicinal plants singly rather than in combinations (266 versus 16 applications) (Grenand *et al.*, 1987).

There are several methods of medicinal plant preparation in Yanomami phytotherapy. For some treatments, e.g. of headaches and stomach aches, pieces of the bark or stems are softened by beating them (to release the juices) and then simply tied around the affected region (Fig. 56). *Renealmia alpinia*, for example, is used in this manner for treating headache. Some of the plants used in this fashion, such as *Uncaria guianensis* and *Aristolochia disticha*, may also be taken internally (as infusions) for the same ailment, suggesting that the active constituents may equally be absorbed through the skin or through the stomach lining. The sniffing or inhalation of crushed aromatic leaves or resins (e.g. *Siparuna guianensis* and *Protium* spp.) is used to treat a number of ailments including congestion, colds, dizziness and nausea. Vapours are also used for treatment of ophthalmia, by heating the leaves or bark in the fire and then holding them up to the open eye (Fig. 60).

The preparation of ingested medicines, many of which are made from the inner bark of trees and vines, varies considerably and may depend upon the circumstances when required. Thus a bark which would be prepared as a hot infusion in the village may, if required in the forest, simply be crushed and squeezed into an impromptu leaf-cup of cold water and drunk (Figs 63, 64). Medicinal barks are generally collected by removing the outer bark from the trunk and then scraping off fine shavings of the inner bark onto a leaf (Fig. 58). These may be wrapped in a tight package made from a marantaceous leaf, and cooked in the embers of the fire (Fig. 62). The cooking releases the juices, which may then be squeezed from the package and drunk.

A few plants and fungi are burned and their ashes used medicinally, particularly for the treatment of oral thrush (*sapinho* in Portuguese) in babies. For this purpose the ashes are glued to the nipple of the nursing mother with

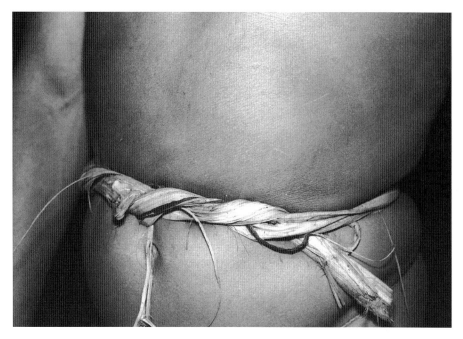

Figure 56. Belt of crushed *Uncaria guianensis* bark for treating stomach ache.

Figure 57. Applying the stinging leaves of *Urera baccifera* to the forehead to relieve headache.

a little saliva or with the sticky sap from the skin of a green plantain, and when the baby is suckled they are effectively dispersed around its mouth (Fig. 61). The dried tannin-rich leaves of *Inga acuminata*, powdered but not burned, are used in the same manner. Both the Wayãpi and the Tiriyó use similar methods of preparation and administration, including the softening of leaves in the fire to release their juices, the burning of leaves to ashes for ingestion or external application, defumation of the afflicted part of the body with heated or burning aromatic plants, rubbing of grated leaves or bark into the skin, application of poultices, etc.

Figure 58. Scraping medicinal bark from a tree trunk.

Figure 59. Blowing the liquid from the stem of a liana, for treating eye infections.

Figure 60. Holding the inner bark of a tree, heated in the fire, to the eyes to treat conjunctivitis.

The proportion of treatments which involve external administration of the plant preparation is high, even for internal disorders. The most prominent are the plants used to treat fevers (e.g. *Piper* and *Peperomia* spp.), most of which are prepared as (aromatic) infusions which are poured over the head and body as a bath. This predominance of external medicine is common in Amazonia; of the 209 preparations recorded among the Tiriyó of Brazil, for example, 41 were applied internally and 168 externally (Cavalcante and Frikel, 1973).

Species employed in Yanomami medicine for their physical rather than for their chemical properties have not been included among the medicinal plants discussed here, but may nonetheless be significant. Arrow cane (*Gynerium sagittatum*), for example, is used for splinting broken limbs, and the red paste from the arils of *Bixa orellana* is applied to the tip of the tongue by mothers to help them to lick foreign bodies out of their babies' eyes. Certain plant species serve as indirect sources of medicine to the Yanomami through their symbiotic relationships with insects. The aggressive *Pseudomyrmex* ants which inhabit the hollow leaf rachises of *Tachigali* aff. *myrmecophila*, for example, are used for medicinal purposes, and the hover-fly larvae (rat-tailed maggots) which inhabit the flooded inflorescence bracts of *Heliconia bihai* are used to remove wax from the ears. These larvae are apparently inserted head-first into the auditory canal, burrow into the ear consuming the wax as they go, and then emerge spontaneously.

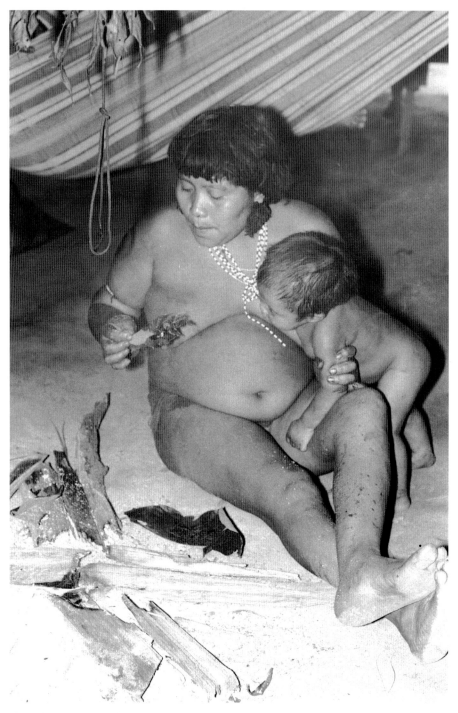

Figure 61. Burning medicinal leaves, whose ashes will be applied to the mother's nipple to treat the child's oral thrush.

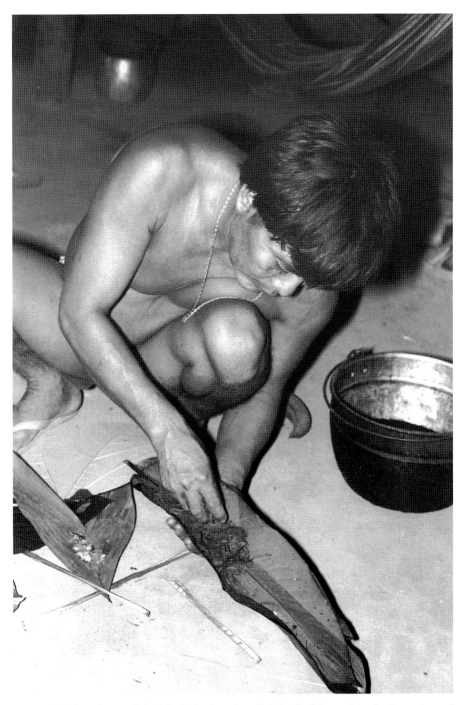

Figure 62. Wrapping medicinal bark shavings in a *Calathea* leaf for cooking in the embers of the fire.

Figure 63. Preparing a cold-water infusion of *Anacardium giganteum* bark for treating diarrhoea.

Figure 64. Drinking a medicinal infusion of aroid roots from a *Calathea*-leaf cup.

Evolution and efficacy of Yanomami medicinal plants

Most of the Yanomami pharmacopoeia has probably been developed through a long process of trial and experimentation with the local flora, in which taste and smell play an important role. Evidence of experimentation with medicinal plants among the Yanomami was observed at **Watoriki** in 1994, where one man reported that he had prepared a mixture of plants normally used as fish poisons and hallucinogens to treat a leishmaniasis-like cutaneous ailment, apparently successfully. This treatment was not based on prior knowledge that the plants would heal the lesions, but on their obvious 'power' as witnessed by their other properties.

Certain tastes are associated with particular medicinal properties by the Yanomami, and to some extent these are used as indicators for the identification of medicinal species. Plants which are peppery to the taste or which sting (*hrami*) are associated with skin disorders such as itching (*xuhuti*). Those which are bitter (*koaimi*) are used to 'kill' unknown pathogenic agents within the body (*yai tʰëpë* or 'invisible, unknown, unnamed things', in this case internal parasites including malaria), and those which taste acidic (*naxi*) are good for regaining energy and taste. The conscious association of bitter-tasting plants with anti-malarial activity is common in the northern Amazon (Milliken, 1997), and some of these other taste-activity associations are probably similarly widespread.

Some medicines, however, may have been learned from neighbouring indigenous groups with whom the Yanomami have traditionally maintained trading links or sporadic contact. Finkers (1986), for example, observed the introduction of *Zingiber officinale* to a village and its rapid spread to other villages in the area after it was recommended by the Ye'kuana as a catarrh remedy. Most of the groups which would in the past have been neighbours of the Yanomami (with the exception of the Ye'kuana) have been extinct for a long time (see Albert, 1985), making analysis of these historical contacts difficult.

Colchester and Lister (1978), during a general ethnobotanical survey in the Orinoco-Ventuari, recorded 16 species of medicinal plants among the Sanima (northern Yanomami) in Venezuela, and 101 species with the neighbouring Ye'kuana. They concluded that: "whereas the Piaroa and Ye'kuana possess very well developed herb-lores, the Macu and Sanema have almost none at all. Moreover, many of the few remedies mentioned by the Sanema [Sanima]are clearly recently learned from the Ye'kuana.....". In the area in which Colchester and Lister were working the Ye'kuana and Sanima had been living in close contact for several decades, and the fact that the latter had not assimilated most of the (diverse) Ye'kuana pharmacopoeia was interpreted by these authors as suggesting that the effect of those plants may have been more symbolic than pharmacological.

There are in fact certain similarities between the pharmacopoeia of the northern Yanomami and the Maiongong (Ye'kuana)[64]. Ten species were collected whose medicinal uses were identical to those in Ye'kuana medicine (e.g. *Anacardium excelsum, Bauhinia guianensis, Dieffenbachia bolivarana* and *Peperomia macrostachya*), and seven which are also used medicinally by the Ye'kuana but for other purposes (e.g. *Renealmia alpinia*). Whether these similarities are due to parallel discoveries of the medicinal properties of the local flora, or to a transfer of information between the groups, would be hard to determine. It may indeed have been a combination of the two[65].

It would not be possible adequately to assess the efficacy of the Yanomami pharmacopoeia without lengthy and detailed pharmacological screening. However, the suggestion that the plants are not directly responsible for the medicinal properties with which they are attributed is questioned by the very considerable correlation which can be found with the medicinal uses to which these plants are put by other peoples in Amazonia (or, in some cases, further afield). In many cases such parallels are found between peoples who are separated by significant linguistic and geographical distances, suggesting that they may have been discovered separately.

---

[64] General collecting of medicinal plants was conducted among the Ye'kuana at Auaris and Uaicas (within the Yanomami territory) by WM in 1994.

[65] Early relationships between the Yanomami and the Ye'kuana were hostile, and only developed into amicable trading links at the turn of the century (Albert, 1985: 40; Colchester, 1982: 87). Whether the intervening period has been sufficient for the transfer of information to the northern Yanomami, and then its subsequent passage to the southeastern Yanomami, is a matter for speculation.

Figure 65. Maiongong (Ye'kuana) women and children at Uaicas on the Rio Uraricoera, within the Yanomami area.

Species:	***Vismia cayennensis* Jacq.) Pers.**
Family:	Guttiferae
Yanomami names:	***yoasi hi, witari mahi*** (Demini)
Brazilian name:	lacre

Description: Common tree in secondary forest at the forest margin. Bark exudes a bright orange sticky latex when cut.

Medicinal use: Remedy for '*pano branco*' skin infections (pityriasis[66]), and for cutaneous leishmaniasis-like lesions (Demini).

Application: For pityriasis, the latex is rubbed on the affected region, ideally after the skin has been abraded with the leaves of *Pourouma bicolor*. For leishmaniasis, the latex is rubbed on the edges of the lesions.

Comparative data: In French Guiana this species is used externally to alleviate itching and to treat wounds, and a decoction of the leaves is drunk to lower fevers (Heckel, 1897). The Wayãpi of that country use the latex from the bark to treat fungal infections of the mouths of children, and the Créoles apply the latex from the fruit to leishmaniasis lesions (Grenand *et al.*, 1987). In Pará (Brazil), the latex is also used for fungal skin disorders (*pano branco*). *V. angusta* is used by the Tikuna Indians to treat skin fungus and herpes of the lips (Schultes and Raffauf, 1990), and is also used for treating skin fungus in Peru (Duke and Vasquez 1994). In Guyana it is used for measles, ulcers, yaws, ringworm and other skin infections (Lachman-White *et al.*, 1987). The barks of many *Vismia* species are rich in quinones, tannins and flavonoids (Grenand *et al.*, 1987).

Figure 66.

All of the medicinal plants listed in Table 8 (p. 88) have been recorded as used elsewhere for the same purposes[67]. The use of *Anacardium giganteum*, *Bauhinia guianensis* and *Uncaria guianensis* for the treatment of diarrhoea and stomach ache, for example, is widespread throughout their ranges and is probably at least partly due to the tannin content of these plants. *Zanthoxylum* bark is used for treating toothache as far afield as Mexico and the USA (Morton, 1981; Schultes and Raffauf, 1990), *Abuta rufescens* is used for treating malaria in Peru (Duke and Vasquez, 1994), and there are numerous records of *Aristolochia* vines being employed in the relief of stomach ache and diarrhoea (e.g. Morton, 1981). These data are discussed in more depth by Milliken and Albert (1996; 1997a).

Malaria, medicinal plants and the Yanomami – acquisition or experiment?

One of the more intriguing questions which arose from the initial medicinal plant studies at *Watoriki* was where (and when) the community's knowledge of anti-malarial plants originated. The development of this knowledge is interesting in that it sheds some light on the wider history of acquisition and development of medicinal plants, and associated 'traditional' knowledge. Since the invasion of their lands by gold prospectors (*garimpeiros*) at the end of the 1980s, malaria has posed a very serious health problem for the Yanomami and has caused a great number of deaths (see Introduction). Given that they had been living in a state of relative geographical isolation until that time[68], it was originally assumed that they would not have suffered significant exposure to the disease, and would consequently possess little knowledge of appropriate medicinal plants. However, it soon became apparent that this was not the case at *Watoriki*, where 10 specifically anti-malarial species were collected (seven used internally and three externally)[69].

Even in regions which had been almost completely isolated from contact with white people prior to the arrival of the *garimpeiros* in 1987, anti-malarial plants were found to be known, suggesting that these populations have been exposed to the disease at some point in the past, contrary to what was originally supposed. At Xitei, for example, four internally administered species were collected, including three which had not been collected at *Watoriki*, and several others which were specified for external use either as baths or as compresses for the enlarged spleen[70].

[66] This manifests itself as white blotches on the skin, often on the shoulders.

[67] They were selected specifically on that basis (see footnote 62).

[68] Apart from the aborted construction of the Perimetral Norte highway across the southern Yanomami lands between 1973 and 1976.

[69] These included *Aspidosperma nitidum* (commonly known as *carapanaúba*), which is widely used for the same purpose by all of the indigenous groups of Brazil's Roraima State (Milliken,1997).

[70] Although there is a strong probability that many of the medicines used externally are treating the symptoms of malaria (fevers, splenomegaly, body pains etc.) rather than the disease itself, the possibility of absorption through the skin of active anti-plasmodial compounds cannot be discounted.

Figure 67. Cutting the bitter bark from a liana used to treat malaria. Antonio is one of the few older men at *Watoriki* who still remembers well the medicinal plants which were used when the group lived in the highlands.

An answer to this enigma, both in terms of exposure to the disease and transmission of anti-malarial knowledge, may lie in the contacts of these groups in the more remote past. The people of **Watoriki** have been in indirect contact (through other Yanomami villages) with neighbouring peoples such as the Ye'kuana and the Makú of the lower Parima since the first decades of the 20th century. The ancestors of the Xitei people in the upper Parima area probably had direct visiting contact with the same groups during the same period, or even before. It may therefore be that they learned of these anti-malarials through these connections. However, a fuller understanding of how and when the knowledge developed would require a more detailed understanding of the history of the disease in the area. A transcription of a statement recorded in 1995 by Roberto, an old headman of Toototobi (since, sadly, deceased), provides some insight:

> 'We got malaria [*hura*: 'spleen-disease'] from the beginning, when we lived over there, in the highlands. People used to travel to the *Hero u* [tributary of the upper Mucajaí] where the *Watatasipë* lived [extinct non-Yanomami Amerindian group]. That's where people got to know malaria. That's where people used to go and visit. Do you know the *Hero u*? Do you know the *Watatasipë* people? These *Watatasipë* were strangers. Our people used to get there from the place called *Konokepë* [an old garden on the upper Toototobi river] to visit the *Watatasipë* and get metal tools [pieces of worn out machetes and axes] from them. They used to go there and get only a few metal pieces. There were also the *Maitʰapë* people [probably another extinct Amerindian group]. Did you know them? The old people first made contact with the *Maitʰapë*. They didn't know you *napëpë* [strangers/ enemies = white people]. The old people first made contact with the *Maitʰapë* from where they lived in the highlands, very far away [between the upper Toototobi and the Orinoco rivers]. It was with these people that they first learned to be contaminated with malaria. So they cured themselves with these things [quoted plant names] because there were no 'white people' at that time. There were no whites and none of their medicines.'

This statement supports our earlier speculation (Milliken and Albert, 1996) that at least some groups of Yanomami had come into contact with malaria earlier than had previously been supposed, perhaps as early as the first decades of the 20th century, and that their knowledge of anti-malarial medicine has been evolving since then. This is further supported by the observations of Holdridge (1933), who noted the presence of endemic malaria on the lower Demini and Aracá rivers in the early 1930s[71], and Smole (1976: 50) who mentioned a probable malaria epidemic in 1935–40 on the Padamo river (Parima highlands), an area inhabited by the Ye'kuana[72].

Given therefore that some Yanomami groups have probably been in sporadic contact with malaria for a very considerable period of time, through visits to infested areas and populations in the past, it is understandable that since their first encounters with the disease they have been developing their knowledge of anti-malarial medicine, either through experimentation or through exchange of information with other peoples.

[71] The middle reaches of these rivers were inhabited by the Bahuana (an Arawak group). White settlers had moved into this region to collect *balata* (latex) and *piaçaba* (palm fibre) on the fringe of the Yanomami territory.

[72] The Yanomami were in contact with a number of neighbouring tribes at the beginning of the century, including the Bahuana, the Purukoto, the Maku and the Ye'kuana, and some may have contracted malaria during that period (see Albert 1985: 59–60).

Plants for fire

The Yanomami traditionally made fire by rotating the point of a hardwood stick (drill) in an indentation in another piece of wood until the friction produced was sufficient to ignite tinder (e.g. the dried pith from the stem of *Gynerium sagittatum*). Hunters and travellers would wrap these sticks (***poro hiki***) in a *Geonoma* leaf and attach them to the bamboo quivers slung across their backs. This fire-making process is no longer used at ***Watoriki*** village and has probably been discontinued almost entirely by the Yanomami. The preferred wood for the fire-drill was *Theobroma cacao*, but *T. subincanum, Bixa orellana, Herrania lemniscata, Guatteria* spp. and *Rinorea lindeniana* could also be used. At Balawaú it was said that a dried section of the stem of the common liana *Bauhinia guianensis* will serve as the base-stick, although it was apparently usual to use the same wood as the drill. The dry leaves of *Geonoma* palms burn well and are good for getting a fire started, and may also be used for smoking out bees, of for burning fleas on the ground in the ***yano***.

Figure 68. Chopping firewood in the garden.

According to Prance (1972a), *Croton pullei* was the principal species used in his study area on the upper Uraricoera river, the soft wood having first been hardened over the fire, and Cocco (1987) mentions the use of *Gossypium barbadense*. *Guatteria* spp. are also used for fire-drills by the Ka'apor and by the Waimiri Atroari (Balée, 1994; Milliken *et al.*, 1992). The inflammable resins exuded by the wounded trunks of certain forest trees, such as *Protium* spp. and *Hymenaea* spp., may be used to help get a fire going in damp conditions or may be burned as torches at night[73]. Such use of inflammable resins (*breu* in Portuguese) is also very common in Amazonia and the Guianas (e.g. Roth, 1924).

Fans for blowing fires aflame are woven from the young (yellow) leaflets of the unopened palm fronds of *Astrocaryum aculeatum* in much the same way that they are made by most Amazonian peoples. Lizot (1984) also records the use of *Jessenia* leaves for this purpose. The Yanomami consume reasonably large quantities of fuel, as fires need to be kept burning all night in each hearth for warmth. Most is collected (by the women) from the unburned felled trees in swidden clearings (Fig. 68). Certain tree species are preferred for firewood, and it was said that the wood of *Anaxagorea acuminata* and *Duguetia* spp., *Elizabetha leiogyne*, *Inga sarmentosa*, and *Sagotia racemosa* burn particularly well. Another species (**waxi hi**), which grows in the highlands, was said by the people of the Catrimani river to be extremely good for burning and is the preferred wood for cooking and cremations.

The high combustibility of the wood of *Sagotia racemosa* (commonly known as *peruano* in Brazil), which is said to burn well when green and thus to be ideal for use while trekking, is widely known among the indigenous groups of Roraima (WM pers. obs.). The Guajá of the eastern Amazon traditionally used it as 'torches' to keep fires burning, and the Ka'apor regard it as one of the best species for toasting cassava (Balée, 1994). Lizot (1984) lists certain species of *Elizabetha*, *Guarea*, *Gustavia*, *Licania*, *Miconia*, *Tabebuia*, *Talisia* and *Touroulia*, an unidentified Sapotaceae (*Micropholis?*) and an unidentified Myrtaceae (*Eugenia?*) as the best for firewood. Cocco (1987) cites *Licania* sp. as one of the best fuels.

Other tree genera cited at **Watoriki** as suitable for firewood include *Acacia*, *Alexa*, *Amphirrhox*, *Aniba*, *Aspidosperma*, *Capirona*, *Casearia*, *Chrysochlamys*, *Chrysophyllum*, *Croton*, *Duroia*, *Ecclinusa*, *Eugenia*, *Fusaea*, *Guarea*, *Guatteria*, *Inga*, *Iryanthera*, *Jacaranda*, *Licania*, *Lindackeria*, *Manilkara*, *Maquira*, *Martiodendron*, *Micropholis*, *Myrcia*, *Ocotea*, *Osteophloem*, *Perebea*, *Picramnia*, *Pogonophora*, *Pouteria*, *Pradosia*, *Protium*, *Quiina*, *Rheedia*, *Rinorea*, *Sclerolobium*, *Sloanea*, *Swartzia*, *Tabernaemontana*, *Tachigali*, *Tetragastris*, *Trichilia*, *Vataireopsis*, *Zizyphus* and *Zollernia*. This list is not exhaustive.

The wood of *Spondias mombin* and *Virola elongata* was said to be used only when nothing better is available, and *Virola* requires removal of the resinous bark before burning. According to Lizot (1984), the wood of *Centrolobium*

---

[73] During **reahu** feasts the collective cassava grating usually goes on during the night, and the resin is burned on the ground to provide illumination.

paraense is avoided as firewood because its smoke is said to attract bats (after which the tree is named by the Yanomami). Several tree species are deliberately avoided as fuel because contact with their ashes causes dermatitis. These include *Bagassa guianensis, Brosimum lactescens, Caryocar villosum, Cordia* cf. *lomatoloba, Helicostylis tomentosa, Pourouma bicolor, P. minor, Pseudolmedia laevigata, P. laevis* and an unidentified Lauraceae (**warë amohi**). It is notable that the majority of these are of the Moraceae family, a number of which contain caustic latex. *Pourouma minor* is also avoided as firewood by the Waimiri Atroari, who maintain that the sawdust causes itching and that smoke from the burning wood causes irritation of the lungs (Milliken *et al.,* 1992). *Pourouma* ash is used as an alkaline admixture to coca in Colombia (Schultes and Raffauf, 1990), and it may be that this alkalinity is responsible for the irritant effect. *Caryocar villosum* is also rejected as firewood by the Ka'apor of the eastern Amazon on account of the itching and eye irritation which it can cause (Balée, 1994). Various species of *Cordia* have been reported to cause dermatitis on account of the cordiachromes (quinoid constituents) which are found in their wood, as have certain alkaloid rich Lauraceae (*Ocotea* spp.) (Hausen, 1981).

Irritant plants

In addition to the species listed above as unsuitable for firewood, a number of other species are also treated with caution by the Yanomami on account of their irritant properties, either chemical or physical. The chippings or sawdust of *Vataireopsis* cf. *surinamensis* are said to sting the skin 'like pepper'. Hausen (1981) also cites *Vataireopsis* as provoking skin irritation and conjunctivitis. Many of the Araceae, such as *Dracontium asperum, Syngonium vellozianum* and *Xanthosoma sagittifolium,* cause itching if their cut surfaces come into contact with the skin, as do the extracts of the *Lonchocarpus* fish-poison vines. These effects are chemical, and have also been reported in these and related genera elsewhere (Behl and Captain, 1979, Benezra *et al.,* 1985). The watera-vine *Pinzona coriacea* is also said to cause irritation – a property which is commonly attributed to several dilleniaceous vines, accounting for their common Brazilian name *cipó de fogo* ('fire-vine'). It is not clear whether the irritation caused by *Cissus erosa* is chemical or physical – possibly the former since the juice of *C. setosa* is also reported to cause dermatitis in India (Behl and Captain, 1979). There are, however, several plants whose reputation is clearly based on mechanical or mechanical/chemical irritation by the hairs or scales found on the leaves, twigs or stems. These include *Guadua* sp., *Herrania lemniscata, Iriartella setigera, Mucuna urens* and *Solanum asperum. Mucuna* acts by injecting (with its hairs) an irritant endopeptidase enzyme into the skin (Behl and Captain, 1979).

Plants in ritual, magic and myth

Shamanic spirits

Of the shamanic spirits (*xapiri*) recognized by the Yanomami, the majority are forest beings, cosmological beings and mythological beings. Among the forest beings, animal spirits (*yaroripë*) predominate, and plants play a relatively minor role[74]. Certain tree species, however, are specifically attributed with shamanic tree spirits (*huu tihipë xapiri* or *huu tihiripë*) by the Yanomami, such as *Virola elongata* and *Tabebuia capitata* whose supernatural forms ('images') are known as curing spirits. The shamanic 'images' of *Anacardium giganteum*, *Cedrelinga catenaeformis*, *Ceiba pentandra*, *Hymenaea parvifolia* and *Vataireopsis* cf. *surinamensis* are also said to have the power of casting out evil spirits. Some of these (e.g. *Ceiba*, *Cedrelinga*, *Hymenaea*) produce some of the largest trees in the forest, and others (e.g. *Tabebuia capitata*) some of the hardest woods, and this may account for their symbolic use in shamanism. *V. surinamensis* is also believed to possess a spirit by the Wayãpi of French Guiana (Grenand *et al.*, 1987).

In addition to the above, generic 'vine spirits' (*thothoxiripë*) and 'leaf spirits' (*yaa hanaripë*) are recognized, which are 'brought down' to prepare the place where a spirit house will be built in the chest of the initiate. These are considered minor spirits in shamanic initiations, and are never individualized. Imaginary trees also play a role in Yanomami shamanism and cosmology. These grow at the end of the world, and include *maa hi* (the rain tree), *amoa hi* (the song tree, where feast and shamanic songs come from) and *thoko hi* (the cough tree, from the furthest land of the white people).

Magic, sorcery and ritual plants

'Magic' plants (i.e. plants which are attributed supernatural powers or effects) play an important part in Yanomami life, and many are cultivated specifically in the gardens or, sometimes, in the clearing around the house (*yano a sipo*). Those with negative (harmful) powers tend to be grown in more discrete locations in the gardens, generally next to fallen and burned tree trunks. Magic plants are used to attract the opposite sex (*thua mamoki*), to bring luck in the hunt (*yaro xiki*), to cause harm to other people (*hwëri kiki*), and for a variety of other purposes (e.g. making children grow, providing courage, facilitating work, etc.). Generally these cultivated species are grown and used either by the men or by the women, those used by the men being carefully avoided by the women and *vice versa*. They may constitute a part of the individual exchanges which take place when one group of Yanomami visits another, and are taken to the new village when a group translocates. Their actions are usually very specific, those for hunting usually being used for one

---

[74] Plants play a role in the lives of the spirits themselves. Female spirits, for example, guard the aromatic vine *Securidaca diversifolia* and wear its leaves and flowers in their ears to attract male spirits. Spirits are also said to eat the flowers of forest trees.

type of game only. The disorders attributed to 'sorcery plants' are usually clearly defined[75]. This plant sorcery accounts for 40–60% of the diagnoses made by Yanomami shamans (Albert, 1985). Plant-related sorcery can be subdivided as follows:

- 'Common sorcery'. This is used between allied (closely linked) villages, for conflicts over theft in the gardens, competition over women, insults, meanness, etc. The plants are used (thrown, blown, applied) as a means of revenge: but they are never meant to result in death and their effects can supposedly be cured by shamanism.
- 'Love sorcery'. In a love conflict between men and women (jealousy, anger over refused or withheld sex, etc.), these plants are used to provoke itching or unpleasant discolouration of the skin.
- 'Prints sorcery'. Earth from a footprint (*mae*) will be collected by an angry member of an allied community, and given away to an enemy of the victim's village who will 'treat' the earth with particular sorcery plants. This is said to cause serious infections in the legs of the victim[76], eventually leading to death.
- 'War sorcery'. Enemy sorcerers (*okapë*) are reputed to creep up on their isolated victims and blow the powder of magical plants over them with a blowgun-like tube. This is said to leave the victims feeling weak and dizzy, whereupon they can then be killed by breaking their bones (arms, legs and spine).

Of the 'magic' plants cultivated at **Watoriki**, the majority were varieties of three species. The first of these was *Cyperus articulatus*, whose aromatic rhizomes are attributed numerous properties. Cultivars include **aroari kiki** or **hwëri kiki yai** (used to cause high fever, loss of consciousness and death), **koamaxi kiki** (used to provoke fever, jaundice and vomiting), **marixi uki** (mixed with *Bixa* and rubbed on babies to make them sleep), **mayëpë xiki** (chewed and rubbed on arrow-head bindings to help kill *Ramphastos* toucans), **ohote kiki** (worn as a necklace to facilitate hard work in the garden), **oko xiki** (used by women to harm men who have mistreated them), **tʰua mamoki** (used by men as a love charm), **waitʰiri kiki** (worn as a necklace or chewed and ingested to give courage), **wakamoxi kiki** (used by men to provoke fever and violent convulsions), **xama mamoki** (mixed with *Bixa* and rubbed on the forehead or grated and ingested to ensure success in the tapir hunt) and **yawë kiki** (mixed with *Bixa* and rubbed in vertical streaks on the bodies of children to speed their growth).

The *Cyperus* plants, which varied greatly in stature, were all collected in a sterile state apart from one, whose inflorescence had been deformed by a fungal infection. Thus, although the identification is reasonably sure, some may represent other (closely related) species such as *Cyperus corymbosus*. The relatively widespread and varied medicinal and magical uses of ergot-infected

---

[75] The names and details of 34 of these plants were recorded during linguistic research at **Watoriki**. The *Cyperus* variety known as **oko xiki**, for example, for use by women, causes high fever, loss of consciousness, yellowing of the eyes, and rigidity in the arms and legs leading to collapse.

[76] The same can be done with a person's tobacco or with the remains of his food (**kanasi**).

Figure 69. One of the many *Cyperus* varieties cultivated for their magical properties. The inflorescence is stunted as a result of fungal infection.

Cyperus articulatus by other indigenous peoples in the Amazon has been researched and discussed in some detail by Plowman *et al.* (1990). Those authors identified the infecting fungus as *Balansia cyperi* and isolated alkaloids from the infected tissues. Given that the infection tends to render these plants sterile, they need to be propagated vegetatively. Lewis *et al.* (1987) reported a very similar use of *Cyperus* species by the Jívaro of Ecuador, where "several sedges were cultivated in most household gardens as highly prized possessions...These plants were traded and sold with great care, they were taken with other important items whenever the family relocated, and they were used largely to treat women and children for specific purposes".

The second species used is *Justicia pectoralis*, whose aromatic leaves contain coumarin. Cultivars used include **ohote hanaki** (worn in the ears by women to facilitate hard work in the garden), **t*h*ua hanaki** (used as love charms by men), **amixi hanaki** (sorcery plant causing fever and crazy thirst) and **puu hanaki**[77] or **romi hanaki** (sorcery plant causing lethal diarrhoea). The complications surrounding the identification of the 'varieties' of *Justicia pectoralis* used by the Yanomami are discussed by Chagnon *et al.* (1971).

The third commonly used species is *Caladium bicolor*[78], which includes **tapra** (used by women to provoke itching in men), **totori mamoki** (mixed with *Bixa* and rubbed on the face to enable a hunter to find *Geochelone* tortoises), **xõwa** (used by women as a love charm) and **tihi kiki** (used to cause fever, delirium and jaguar-like behaviour). In addition, *Alstroemeria* sp. is used by men to bring about sterility in women, *Maranta arundinacea* by women to take away men's courage, *Dioscorea* cf. *piperifolia* as a love charm to attract the opposite sex, and the leaves of the cultivated fish poison plant *Clibadium sylvestre* are said to have been burned in the past to provoke epidemics among Yanomami enemies.

There are a great many more varieties of the cultivated 'magic' species described above than are given here, some of which are mentioned by Albert (1985), Albert and Gomez (1997), Eguillor Garcia (1984) and Fuentes (1980). The information regarding these varieties and their properties is likely to vary considerably from area to area. It is interesting to note that the Wayãpi and Palikur Indians of French Guiana also recognize numerous varieties of *Caladium bicolor* (Grenand *et al.*, 1987) and attribute them with a broad range of magical properties (including hunting magic), and that Ka'apor men of the eastern Amazon use *Justicia* leaves to attract women (Balée, 1994).

Numerous wild forest plants are also attributed 'magical' properties by the Yanomami or are used in specific rituals[79]. The leaves of *Psychotria ulviformis*, whose name translates as 'close-leaf', is worn in the armband or behind the

---

[77] This is different from the plant worn by women in their arm bands.

[78] This species was identified from sterile collections. It may be that some of the small variegated aroids cultivated by the Yanomami belong to a different species.

[79] Fuentes (1980) cites a number of wild species as being used for unspecified magical/ritual purposes by the Yanomami which were not recorded as used in this way at **Watoriki** village, including *Amyris* sp., *Apeiba* sp., *Cassia reticulata*, *Costus* sp., *Ficus* cf. *paraensis*, *Geonoma* cf. *baculifera*, *Mesechites trifida*, *Psidium* sp. and *Tachigali paniculata*.

Figure 70. Magical varieties of *Caladium bicolor*. The larger one (**tapra**) causes itching, and the smaller one (**xõwa**) is a love-charm.

ears to make a journey seem shorter. A ritual for ensuring success in a tapir (*Tapirus terrestris*) hunt involves passing the hunting dogs through hoops made of certain vines including *Arrabidaea* sp., *Callichlamys latifolia, Machaerium macrophyllum, Philodendron hylaeae* and *Uncaria guianensis*. Branches of *Sorocea muriculata*, with their vivid crimson peduncles, are used to construct the enclosure for seclusion of young girls at their first menses[80].

Several wild and generally aromatic species are employed as love charms[81] (sometimes in combination), including the pith of the twigs of *Aniba riparia*, the interior of the ripe fruits of *Solanum oocarpum*, the coumarin-rich seeds of *Dipteryx odorata*, the crushed dried seeds of *Picramnia spruceana*, and the exudates from the stems of *Humiria balsamifera* and *Securidaca diversifolia*. According to Fuentes (1980), *Myroxylon balsamum* and *Dorstenia* sp. are also used in this way. The genera *Aniba, Dipteryx, Myroxylon* and *Humiria* have all been commercially exploited for their aromatic properties, the last of these also being used by the Indians of Guyana to anoint their hair (Fanshawe, 1948). The use of aromatic plants by the Yanomami for love-charms has been discussed further by Alès (1987).

The stems of at least three unrelated species of vines with distinctly phallic flowers or fruits, all known by the name of ***ihuru tʰotʰo*** (***ihuru*** = child), are used in fertility rituals by women. These include *Aristolochia* sp., *Passiflora fuchsiiflora* and *Gurania spinulosa*. Conversely, the parasitic *Sciaphila purpurea* is used by men, in conjunction with other plants and sometimes with 'cuckoo-spit' insect larvae, to bring about loss of fertility in hostile women. In the same way that **tapra** (cultivated *Caladium bicolor*) is also used by women in sorcery to provoke itching in hostile men (and eventually in enemies by throwing it on the path to the **yano**), so also are certain wild Araceae including *Syngonium vellozianum* and *Xanthosoma sagittifolium*. The cut surfaces of these species cause actual itching on contact with the skin, as do many Araceae on account of the calcium oxalate crystals typically found in their tissues (see above).

Both wild and cultivated food plants may also be subject to prohibitions at certain times and in certain circumstances. Women, for example, must avoid the toasted seeds of *Posadaea sphaerocarpa* when pregnant or their babies will contract oral thrush, and if pawpaws, yams or *Caryocar* fruits are eaten during rituals associated with bleeding (menstruation, birth, homicide), they will cause inguinal hernias[82].

---

[80] This species was not seen in the vicinity of **Watoriki**, although it is very abundant in the highlands. Another species, *Maprounea guianensis*, was collected there under the same name, but it is not clear whether or not this is used as a substitute for the same purpose. These enclosures now tend more often to be made from banana leaves.

[81] Love charms of this type are specific to the sex, most being used by men to attract women, and some *vice versa*.

[82] There are also prohibitions on the uses of or contact with other (non-edible) plants. This is dealt with in more detail by Fuentes (1980), who reports (for example) that children are told to avoid touching the scarlet flowers of *Psychotria poeppigiana* as this is said to cause them to fall from their hammocks at night and burn themselves in the fire.

Figure 71. *Passiflora fuchsiiflora* – a forest vine with phallic flowers and magical properties.

Plants in myth

As one might expect of any forest people, plants feature prominently in the myths of the Yanomami. These myths have been documented in great detail by several of the anthropologists who have worked among them, and have been compiled as a book by Wilbert and Simoneau (1990), including stories from three principal Yanomami linguistic sub-groups (western and eastern Yanomami, and Sanima). Among these myths are 63 which were collected by BA among the southeastern Yanomami at Catrimani, Toototobi and **Watoriki**, which will be discussed in some detail here[83]. These contain specific references to over 55 species of plants.

In approximately half of the specific references the plants' roles are relatively minor. They form part of the background information of the story without the apparent attachment of any particular symbolism, generally reflecting everyday or incidental uses of the plants or filling in visual details such as the tree which a character is sitting in, the fruit which he is eating or the body paint or fragrant leaves which she is wearing, etc. Some examples of these references are given in Table 11.

However, amongst the plants occurring as such background information are a number of 'prototypic' species whose appearance is recurrent in Yanomami myths, emphasizing their archetypal role in a particular activity. When reference to firewood is made, for example, the species mentioned is almost invariably *Elizabetha princeps* or **waxi hi** (trees which burn particularly well), and when people are described as gathering fruits in the forest these fruits are generally *Pseudolmedia laevigata* or *Micrandra rossiana*, which were traditionally important in the Yanomami diet (see above). Some of these 'prototypic' species of traditional importance (e.g. *M. rossiana* and **waxi hi**) have retained their place in the myths of the lowland Yanomami in spite of the fact that they are absent from the regions in which those people are now living.

In a number of more significant references to plants among the southeastern Yanomami myths, there is a clear symbolic association between the characteristics of the species mentioned and its role in the story. These references clearly provide a key analogy necessary to the narration of the myth. For example, in one creation myth [7] the sky falls onto the crown of a cocoa tree (*Theobroma cacao*), leaving the Yanomami ancestors trapped beneath. They escape into the world above through a hole cut in the sky by a mealy parrot (*Amazona farinosa*), and from then on they live on its surface. The old sky, which now forms the ground, remains deformed by the crown of the cocoa tree over which it is draped, giving rise to the hills of the Serra Parima (in the heart of Yanomami territory), where the rivers are born. The association may also be based on knowledge of the ecological relationships of the plant with animals (myth characters), or on its useful properties. Thus in the story of the origin of fire [50], the crested oropendola (*Psarocolius*

---

[83] Numbers in square brackets in the text and tables below refer to the myth number in Wilbert and Simoneau (1990).

Table 11

Some examples of plants providing background information in eastern Yanomami mythology

Myth	Use in narrative	Role
Omama creates present Yanomami [11]	*Heliconia bihai*	Leaves placed on ground to catch "ants' eggs"
Flood and origin of foreigners [33]	*Virola theiodora*	Men taking **yãkoana** snuff at end of the **reahu** feast
Return of ghosts and origin of death [35]	*Phenakospermum guyannense*	Leaves gathered to make shelters
Toad festival and origin of ceremonial songs [41]	*Micrandra rossiana*	Ceremonial food at festival
Bat, the incestuous son-in-law [42]	*Bixa orellana*	Guests paint themselves with the fruits before the feast
Child warrior [47]	*Strychnos* sp.	Curare arrow heads prepared to avenge mother
Tapir causes hummingbird to be burned [171]	*Elizabetha princeps*	Tapir uses the wood to make pyre for hummingbird
Omama and **Yoasi** take revenge on the jaguar [187]	*Pseudolmedia*	Twins climb the trunk to escape jaguar, and pretend to pick its fruits
Omama and **Yoasi** catch **Tëpërësiki**'s daughter [197]	*Justicia pectoralis*	Daughter decorated with its aromatic leaves
Cannibal husband [246]	*Capsicum frutescens*	Peppers thrown in the fire to smoke the cannibal out of the cave
Western Yanomami boy falls into underworld [277]	*Clathrotopis macrocarpa*	Boys' mothers are washing the seeds in the river
Man transformed into bees' nest [308]	*Oenocarpus bacaba*	Leaves burned to smoke bees out of their nest
Enemy sorcerers fall into precipice [359]	*Iriartella setigera*	Man uses the stem to feel for edge of precipice

decumanus) steals it from the caiman (*Caiman crocodilus*) and flies into the crown of an *Elizabetha princeps* tree. The fire is deposited in the heart of this tree (one of the best for firewood), and likewise into *Herrania lemniscata, Bixa orellana* and *Guatteria* spp. (whose wood is used for fire-drills) and into the arrow cane (*Gynerium saggitatum*) whose pith is used as tinder for fire-making. Further examples of these simple associations are given in Table 12.

In some cases the symbolic reference is more complex, and the plants are used to metaphorize the central social/cultural concept or comparison which the story is attempting to convey. These are relatively few, since the characteristics and behaviour of animals provides a richer source of symbolism. However, in the story of 'red brocket and the false peach palm' [102], for example, an opposition is made between the delicious fruits of

Bactris gasipaes (owned by silver-beaked tanager, the son-in-law) and the barely edible fruits of *Socratea exorrhiza* (owned by red brocket, the father-in-law), which is intended to symbolize the ambivalence of in-law relationships in Yanomami society. Likewise the opposition between mortality and immortality and between good and bad are symbolized through plants in some of the stories about **Omama**, creator of the present Yanomami humanity and society, and **Yoasi**, his indolent and wayward twin brother. In one such myth [191] **Omama** sends **Yoasi** to cut stakes of **pore hi** (a myrtaceous tree with strong wood and constantly peeling bark) to support the hammock of the river monster **Tëpërësiki**'s daughter, whom they caught in the river. His intention is that this

Table 12
Some examples of plants providing basic symbolic associations in eastern Yanomami mythology

Myth	Plant	Use in narrative	Probable reference
Omama creates present Yanomami [11]	*Iriartella setigera*	Objects like ants' eggs are taken from the split stem and turned into Yanomami	Black ants (*Camponotus* sp.) make their nests in this palm
Toad festival and origin of ceremonial songs [41]	*Brosimum utile*	Arm-band of **Pokohayumëri** monster (left-handed giant) compared in joke with bast fibre sling	This is one of the most commonly used species for baby-carrying slings
Opossum and the origin of sorcery [130]	*Tabebuia capitata*	Opossum hides in the tree's hollow trunk	Opossums are often found in these large hollow trees
Enemy sorcerers turn into coatis [141]	*Caryocar glabrum*[84]	Sorcerers eat seeds in ritual state of homicide; their noses are deformed by the fruits as they eat	Coatis eat *Caryocar* fruits; turned up noses of coatis
Tapir tries to hide [172]	*Cecropia sciadophylla*	Tapir and sloth try to hide in the tree; tapir is not successful and goes away; sloth remains (and eats leaves)	Sloths live in trees, are hard to spot, and *Cecropia* leaves are important in their diet
Omama creates mountains in his flight [211]	*Oenocarpus bacaba*	**Omama** throws leaves behind him to hide his tracks, and they become mountains	The fallen leaves look like mountain ridges
Teremë, girl with a claw [251]	*Pseudolmedia laevis*	Yellow-green grosbeak cuts fruit from the tree; its red fruits are confused with **Tërëmë**'s victim's blood	Grosbeaks eats *Pseudolmedia* fruits; the fruits are blood red and squashy
Agouti, the stingy mother-in-law [289]	*Clathrotropis macrocarpa*	Agouti poisons her son-in-law with unprocessed **wapu** fruits	Agoutis are virtually the only animals to eat poisonous **wapu** seeds, which they hide in the ground

[84] See footnote 9

would enable her to change her skin when she became old, the Yanomami thus becoming immortal. The lazy *Yoasi*, however, cuts softer *Croton* wood instead, which does not shed and regenerate its bark like *pore hi*, and the Yanomami are mortal as a consequence. In a similar myth [197], when *Omama* goes to gather sweet-smelling *Ormosia* to rub on *Tëpërësiki*'s daughter's vulva, the impatient *Yoasi* picks the pungent flowers from a nearby *Inga acuminata* tree and uses these instead, explaining the natural scent of women.

There are also several myths in which plants are the main subject of the story, such as those which recount the origins of cultivated species, many of which were brought to the Yanomami by mythical characters. Some of these plants must have been introduced at some indefinable point in the past (see Plants for Food), and these myths perhaps refer obliquely to those events. In the story of the lost mother-in-law [86], a huge field of maize is cultivated by her son-in-law the *Atta* leaf-cutter ant, an insect which eats most of the leaves in the Yanomami gardens but not those of the maize plants. Whereas this narrative could support the argument that maize formed an important part of the diet in the past (see also Colchester, 1984; Valero, 1969, 1984), the story of the origin of the plantain [84], in which the bird people steal it from the garden of a ghost (*pore*), could point to its acquisition from foreigners[85]. In another myth [198][86] the cultivated plants were all obtained from *Tëpërësiki*, the aquatic monster who brought them to *Omama*, his son-in-law, in order that he should be able to feed his daughter. Some myths also tell of the domestication of wild forest species. For example, according to one version [92] of the origin of cultivated yams (*Dioscorea trifida*)[87], they were called from the forest by an old woman who was alone and hungry in the *yano*. When they came to the door she caught, cooked and ate them, and set aside the smallest ones to plant.

Other myths tell of the discovery of the useful properties of wild species. In the story of the caterpillar's tobacco [76], for example, the Yanomami learn about the processing of forest seeds. The caterpillar[88] travels through the forest distributing tobacco to those who are generous to him and punishing those who are not. He encounters the kinkajou people (*Potus flavus*), feeding in an *Inga* tree, and in return for the ripe fruits thrown down to him he teaches them that the seeds are edible if they are boiled. He then teaches the yellow-green grosbeak people (*Caryothraustes canadensis*) how to roast the bitter kernels of *Pseudolmedia laevigata* fruits, the blue-headed parrot people

[85] The first white people encountered by the Yanomami were invariably seen as returning ghosts.

[86] Some myths of origin are duplicated in Yanomami mythology, the first series telling of the *yaroripë* (the first humanity of ancestors/animal people) and the second telling of the 'new order' installed by *Omama* after the fall of the sky. In the stories of the first ancestors, people are frequently transformed into animals as punishment for their improper actions (which generally invert today's proper social behaviour).

[87] This species is cultivated by the Yanomami but is also found growing wild in the forest.

[88] *Manduca sexta* (*Yoropori*), a voracious eater of tobacco leaves but also the bringer of tobacco to the Yanomami.

(*Pionus menstruus*) how to boil the seeds of another species of *Inga*, and so on. A considerable amount of information about ecological relationships (e.g. plant-animal interactions) is obviously embedded within many of these myths.

Considering the body of Yanomami myths as a whole, there is a common core represented throughout the linguistic sub-groups, with slight or sometimes significant variation in their details, as well as myths which are particular to each sub-group. But even when the story is virtually identical the plants or animals mentioned may be different, as long as they fulfil the same symbolic purpose. For example, in another version of the fire origin story recorded by Lizot among the western Yanomami, the same bird steals the fire but he carries it into a *Micropholis* tree (which is also good for fuel), and in a Sanɨma version the long-tailed tyrant (*Colonia coloniis*) steals it and carries it into the crown of a *Coussapoa* tree (used by that sub-group for fire-drills). Such differences may also be found amongst the myths told by communities of the same sub-group, or even those told by individuals within the same *yano*.

Some parallels in plant symbolism can also be found between the myths of the Yanomami and those of other Amazonian indigenous peoples. For example, the role of lianas in the passage of mythical characters or spirits from the earth to the sky or *vice versa* is common in Amazonian folk literature. According to the southeastern Yanomami [35] the 'ghosts' of the dead used to commute regularly between one and the other, spending periods of time among the living, until a mealy parrot cut the *Clusia* vines which they used to use as a ladder, stranding them in the sky. According to the Sanɨma [145] all the animals were climbing a *Bauhinia* vine when it finally broke under the weight of the tapir. Most of the animals fell to earth, but the monkeys, who had climbed higher than the others, fell into the forest canopy and remained there. This is a remarkably similar story to one told by the Waimiri Atroari (see Milliken *et al.*, 1992) and indeed by most if not all Carib speaking tribes, and the choice of *Bauhinia* as the 'ladder' in all these stories clearly owes its origin to the curiously distorted stems of many members of that genus[89].

YANOMAMI ETHNOBOTANY IN PERSPECTIVE

Lizot (1984) reported that the Yanomami with whom he worked in the Orinoco recognized 328 plant species, of which they used 57%. It is now clear that this was a very considerable underestimate, both of the botanical knowledge of the Yanomami and of the number of species used by them; the number of medicinal species recorded to date among the Brazilian Yanomami alone exceeds this level. In fact, the data necessary to make a reasonably accurate assessment of the numbers of species known to and used by the Yanomami are still not available, and those presented here only represent the results of a necessarily partial and unbalanced study. Many more years of intensive research would be required to document this information fully, and

[89] *Bauhinia* vines are commonly known as *escada de jabuti* (tortoise ladder) in the Brazilian Amazon.

it is arguable that it could only then be done properly by the Yanomami themselves. Although the data presented in this paper touch upon most aspects of Yanomami plant use, and in some cases these have been researched in some depth, it is clear that there are large tracts of Yanomami ethnobotany, such as their knowledge of ecological interactions involving plants, which have still scarcely been touched upon.

In this publication, the numbers of species used by the Yanomami for particular purposes are compared with those recorded amongst other indigenous peoples of the Amazon, and in most cases they are shown to use as many as (or more than) most of the groups with whom they are compared. For the case of the medicinal plants, for example, which was the most intensively surveyed category during the present study (and therefore perhaps the most representative), the highest recorded number of species among the comparative literature studied was 245 for the Shuar of Ecuador (Bennett, 1992), whereas the majority reported very substantially fewer than the 198 reported here (see p. 87). However, comparisons of the levels of plant knowledge and use between indigenous peoples are inevitably skewed by the differing (and generally unquantifiable) intensities of studies which they have undergone, and although these inequalities can to some extent be overcome by limited-area quantitative surveys, such as the single-hectare forest studies described by Prance *et al.* (1987) and subsequent authors (e.g. Milliken *et al.*, 1992), the extent to which these permit meaningful comparisons is still limited. Furthermore, even if balanced comparisons were possible, their results would be of dubious significance in the wider cultural and ecological context.

It is more informative to compare the ethnobotany at a qualitative level, where it has been shown that the similarities in the use of species between the Yanomami and other indigenous peoples of the Amazon are extremely high. This level of correlation is not surprising in some respects, given that these peoples have been living for long periods[90] in basically the same environment and with the same levels of technology available to them. But, if one considers the extreme diversity of the Amazonian flora, i.e. the vast numbers of species from which they have had to choose for any one application, coupled with the linguistic barriers and the massive geographical distances involved, it is perhaps more remarkable.

If the numbers of species used by the Yanomami are unusually high, this is partly due to the ecological variations found within the Yanomami area and the diversity of knowledge acquired by different sub-groups of the people themselves. At the ecological level, the territory which they currently inhabit is extremely diverse. It spans a broad altitude range (100 m to 1600 m), including summit and outcrop floras, highland savannas (in the Serra Parima), montane cloud forests, submontane forests and lowland floodplain forests (Huber *et al.*, 1984). The range of species available to the Yanomami is

---

[90] Recent archeological discoveries point to human occupation of the Amazon region for at least 10,000 years (Roosevelt *et al.*, 1996).

therefore very great, perhaps exceptionally so. In addition, the very dynamic process of village fission and geographical expansion experienced by the Yanomami since the 19th century has probably permitted them to experiment with macro- and micro-local floristic diversity to an unusual degree. However, although the incomplete nature of the data still precludes detailed analysis of these differences, it appears that the use of plants by the lowland Yanomami (*Watoriki*, Balawaú) is essentially very similar to that of the highland Yanomami (Xitei), where the same species are available for use. Likewise, there are few fundamental ethnobotanical differences (discounting plant names) between the Yanomami studied by the authors and those reported by other researchers in Venezuela (e.g. Fuentes, 1980; Lizot, 1984).

Nevertheless, the fact that these ethnobotanical variations are limited does not imply identical use of species by the different Yanomami sub-groups. The Yanomami territory has expanded considerably since the 19th century (see Albert, 1985), and many of the groups now inhabiting the peripheral lowland regions, including the *Watoriki* people discussed here, have only been living in those regions for a relatively short time (about half a century). The Brazilian and Venezuelan groups have moved down from a shared highland 'homeland' (see Introduction), and have been in the process of adapting, through their ethnobotany and through other means, to new environments and new situations for several decades.

The forests on the Venezuelan side of the Serra Parima represent the same basic Guayanan-Amazonian flora as those on the Brazilian side, but there will inevitably be some local and regional differences between them, both in their overall composition and in the relative abundance of their component species. As the Yanomami have moved down into these forests they have gradually adapted their use of species to suit the available resources, without fundamentally altering their ethnobotany. This process has probably been facilitated by the relatively high diversity plants which they used before they migrated, raising the probability of their encountering at least one of the species required for a particular purpose in their new environment.

In many cases this adaptation has meant employing new members of the genera which were originally employed in the highlands or, more rarely, related (or sometimes unrelated) genera with similar properties. The *Rinorea* species used for arrow nocks by the *Watoriki* people, for example, is different from that reported among the Venezuelan Yanomami by Fuentes (1980), but that difference is probably a reflection of nothing more than the relative availabilities of different *Rinorea* species in the two areas.

The changes experienced by the Yanomami are not purely environmental, but also stem from the rapidly escalating contact with the outside world which they have experienced over the last decades, the phenomena of increasing sedentariness, and, in some cases, escalating dependence on externally supplied resources such as medicines. The significant effects of these changes on their use of plants for house construction and on their use of medicinal plants have been discussed in some depth (Milliken and Albert, 1996, 1997a, 1997b). The Yanomami, however, are a highly curious, receptive and adaptable people, open to new technologies and resources, and no aspect of their material culture is written in stone. Their ethnobotany, which doubtless already bears the stamp of previous changes in their history, whether geographical or otherwise, and of their contacts with other peoples (indigenous or not), is a dynamic phenomenon.

ACKNOWLEDGEMENTS

None of this work would have been possible without the co-operation, hospitality and kindness of the Yanomami people of *Watoriki* village, particularly Antonio, Justino, Lourival, Davi, Roberto, Lucas and Henrique, and likewise the people of the other communities which were visited. The research was carried out with the authorization of the Fundação Nacional do Índio (FUNAI), under the aegis of the Conselho Nacional de Desenvolvimento Tecnológico e Científico (CNPq) and the Universidade de Brasília (UnB), in collaboration with Profª Alcida Ramos and the Museu Integrado de Roraima (MIRR). It was funded and supported by the Royal Botanic Gardens, Kew, the Baring Foundation, the Ernest Cook Trust, the Rufford Foundation, George Mark Klabin and the Rainforest Medical Foundation. The staff of CCPY (Commissão Pró-Yanomami) in São Paulo, Boa Vista and *Watoriki*, and Irmã Ninfa of the Catholic mission at Xitei, gave invaluable support and assistance. Ana Paula Souto Maior is also gratefully acknowledged, as are Suzy Dickerson and Christine Beard for their long-suffering editorial support. The Royal Botanic Garden Edinburgh provided facilities during the final stages of writing, and Hew Prendergast, Ghillean Prance and Jim Ratter made valuable comments on the draft manuscript.

LITERATURE CITED

Acevedo-Rodríguez, P. (1990). The occurrence of piscicides and stupefactants in the plant kingdom. *Advances in Economic Botany* 8: 1–23.

Aguiar, J.P.C., Marinha, H.A. Rebelo, Y.S. and Shrimpton, R. (1980). Aspectos nutritivos de algunas frutas da Amazônia. *Acta Amazonica* 10 (4): 755–758.

Agurell, S., Holmstedt, B. and Lindgren, J.-E. (1969). Alkaloids in certain species of *Virola* and other South American plants of ethnopharmacologic interest. *Acta Chemica Scandinavica* 23: 903–916.

Albert, B. (1985). *Temps du sang, temps des cendres. Représentation de la maladie, système rituel et espace politique chez les Yanomami du sud-est.* Doctoral thesis, Université de Paris X-Nanterre.

Albert, B. (1990). On Yanomami warfare: a rejoinder. *Current Anthropology* 31: 558–562.

Albert, B. (1994). Gold miners and Yanomami Indians in the Brazilian Amazon: the Haximu massacre. Pages 47–55 *in* B.R. Johnston, ed., *Who pays the price? The sociocultural context of environmental crisis.* Island Press, Washington.

Albert, B. and Goodwin Gomez, G. (1997). *Saúde Yanomami. Um manual etno-lingüístico.* Museu Paraense Emílio Goeldi, Belém.

Alès, C. (1987). Les parfums végétaux des Yanomamɨ (indiens du Venezuela). Pages 242–246 *in Parfums de Plantes.* Museum National d'Histoire Naturelle, Paris.

Altschul, S. von R. (1972). *The genus* Anadenanthera *in Amerindian cultures.* Harvard University Press, Cambridge.

Altschul, S. von R. (1973). *Drugs and foods from little-known plants.* Harvard University Press, Cambridge.

Anderson, A.B. (1978). The names and uses of palms among a tribe of Yanomama Indians. *Principes* 22: 30–41.

Balée, W.L. (1987). A etnobotânica quantitativa dos Índios Tembé. *Boletim do Museu Paraense Emílio Goeldi, Série Botânica* 3: 29–50.

Balée, W.L. (1989). The culture of Amazonian forests. *Advances in Economic Botany* 7: 1–21.

Balée, W.L. (1994). *Footprints of the forest. Ka'apor Ethnobotany – the historical ecology of plant utilization by an Amazon people.* Columbia University Press, New York.

Balick, M.J. and Gershoff, S.N. (1981). Nutritional evaluation of the *Jessenia bataua* palm: source of high quality protein and oil from tropical America. *Economic Botany* 35 (3): 261–271.

Barfod, A.S. and Kvist, L.P. (1996). Comparative ethnobotanical studies of the Amerindian groups in coastal Ecuador. *Biologiske Skrifter* 46: 1–166.

Beck, H.T. and Prance, G.T. (1991). Ethnobotanical notes on Marajó ceramic pottery utilizing two Amazonian trees. *Boletim do Museu Paraense Emílio Goeldi, Série Botânica* 7 (2): 269–275.

Behl, P.N. and Captain, R.M. (1979). *Skin-irritant and sensitizing plants found in India.* S. Chand & Co., New Delhi.

Benezra, C., Ducombs, G., Sell, Y. and Foussereau, J. (1985). *Plant contact dermatitis.* C.V. Mosby, London.

Bennett, B.C. (1992). Plants and people of the Amazonian rainforests. The role of ethnobotany in sustainable development. *BioScience* 42 (8): 599–607.

Bennett, B.C. and Andrade, P.G. (1991). Variación de los nombres y de los usos que dan a las plantas los indígenas Shuar del Ecuador. Pages 129–137 *in* M. Ríos and H. Pedersen eds., *Las plantas y el hombre.* Ediciones Abya-Yala, Quito.

Biocca, E. (1979). Sciamanismo, allucinogeni e meloterapia: relazione introduttiva. Pages 445–453 *in Simposio internazionale sulla medicina indigena e populare dell' America Latina,* IILA-CISO Rome 12–16 December 1977. IILA, Rome.

Boom, B.M. (1987). Ethnobotany of the Chácobo Indians, Beni, Bolivia. *Advances in Economic Botany* 4: 1–68.

Boom, B.M. (1990). Useful plants of the Panare Indians of Venezuelan Guayana. *Advances in Economic Botany* 8: 57–76.

Borgman, D. (1990). Sanuma. Page 147 *in* D.C. Derbyshire and G.K. Pullum eds., *Handbook of Amazonian Languages.* Mouton de Gruyter, New York.

Branch, L.C. and Silva, M.F. da (1983). Folk medicine of Alter do Chão, Pará, Brazil. *Acta Amazonica* 13 (5–6): 737–797.

Brewer-Carias, C. and Steyermark, J.A. (1976). Hallucinogenic snuff drugs of the Yanomamö Caburiwe-Teri in the Cauaburi River, Brazil. *Economic Botany* 30 (1): 57–66.

Carneiro, R.L. (1979a). Tree felling with the stone ax: an experiment carried out among the Yanomamö Indians of Southern Venezuela. Pages 21–58 in C. Kramer ed., *Ethnoarchaeology: Implications of ethnography for archaeology.* Columbia University Press, New York.

Carneiro, R.L. (1979b). Forest clearance among the Yanomamö, observations and implications. *Antropológica* 52: 39–76.

Cavalcante, P.B. and Frikel, P. (1973). A farmacopeia Tiriyó. Museu Paraense Emílio Goeldi, Belém.

Chagnon, N.A. (1966). *Yanomamö warfare, social organization, and marriage alliances.* Doctoral thesis, University of Michigan.

Chagnon, N.A. (1968). *Yanomamö, the fierce people.* Rinehart, Holt and Winston, New York.

Chagnon, N.A. (1974). *Studying the Yanomamö.* Holt, Rinehart and Winston, New York.

Chagnon, N.A., Le Quesne, P. and Cook, J. (1970). Algunos aspectos de uso de drogas, comercio y domesticación de plantas entre los indígenas yanomamö de Venezuela y Brasil. *Acta Científica Venezolana* 21: 186–193.

Chagnon, N.A., Le Quesne, P. and Cook, J. (1971). Yanomamö hallucinogens: anthropological, botanical, and chemical findings. *Current Anthropology* 12 (1): 72–74.

Chiara, V. (1987). Armas: bases para uma classificação. Pages 117–138 *in* D. Ribeiro ed., *Suma etnológica Brasileira.* Vol. 2: Tecnologia indígena. Vôzes, Petrópolis.

Cocco, L. (1987). *Iyëwei-theri. Quince años entre los Yanomamos.* 2nd Edition. Escuela Técnica Don Bosco, Caracas.

Colchester, M. (1982). *The economy, ecology and ethnobiology of the Sanema Indians of South Venezuela.* Doctoral Dissertation, University of Oxford.

Colchester, M. (1984). Rethinking stone age economics: some speculations concerning the Pre-Colombian Yanoama economy. *Human Ecology* 12 (3): 291–314.

Colchester, M., and Lister, J.R. 1978. *The ethnobotany of the Orinoco-Ventuari region.* An introductory survey. Unpublished ms.

Czerson, H., Bohlmann, F., Stuessy, F. and Fischer, H. (1979). Sesquiterpenoid and acetylenic constituents of seven *Clibadium* species. *Phytochemistry* 18: 257–260.

Davis, E.W. and Yost, J.A. (1983). The ethnobotany of the Waorani of Eastern Ecuador. *Botanical Museum Leaflets, Harvard University* 29 (3): 273–297.

Denevan, W.M. (1992). Stone vs metal axes: the ambiguity of shifting cultivation in prehistoric Amazonia. *Journal of the Steward Anthropological Society* 20: 153–165.

Denevan, W.M. and Treacy, J.M. (1987). Young managed fallows at Brillo Nuevo. *Advances in Economic Botany* 5: 8–46.

DSY/RR – FNS (1999). *População Yanomami por pólo-base e comunidades. Distrito Sanitário Yanomami/Roraima* – Fundação Nacional de Saúde, Boa Vista.

Duke, J.A. and Vasquez, R. (1994). *Amazonian ethnobotanical dictionary.* CRC Press, Boca Raton.

Eguillor Garcia, M.I. (1984). *Yopo, shamanes y hekura. Aspectos fenomenológicos del mundo sagrado Yanomami.* Libreria Editorial Salesiana, Vicariato Apostólico de Puerto Ayacucho.

Fanshawe, D.B. (1948). Forest products of British Guiana. Part II. Minor forest products. *Forestry Bulletin* 2 (new series): 7–81.

Fidalgo, O. (1965). Conhecimento micológico dos Índios Brasileiros. *Rickia* 2: 1–10.

Fidalgo, O. and Hirata, J.M. (1979). Etnomicologia Caiabi, Txicão e Txucarramãe. *Rickia* 8: 1–5.

Fidalgo, O. and Prance, G.T. (1976). The ethnomycology of the Sanama Indians. *Mycologia* 68 (1): 201–210.

Finkers, J. (1986). *Los Yanomami y su sistema alimentício.* Vicariato Apostólico de Puerto Acacucho. Monografía No. 2.

Flores Paitán, S. (1987). Old managed fallows at Brillo Nuevo. *Advances in Economic Botany* 5: 53–66.

Frechione, J., Posey, D.A. and Silva, L.F. da. (1989). The perception of ecological zones and natural resources in the Brazilian Amazon: an ethnoecology of Lake Coari. *Advances in Economic Botany* 7: 260–282.

Fuentes, E. (1980). Los Yanomami y las plantas silvestres. *Antropológica* 54: 3–138.

Fuerst, R. (1967). Die gemeinschaftswohnung der Xiriana am Rio Toototobi (Beitrag zur kenntis der Yanomami-Indianer in Brasilien). *Zeitschrift Für Ethnologie* 92 (1): 103–113.

Fukami, J., Yamamoto, I and Casida, J.E. (1967). Metabolism of rotenone *in vitro* by tissue homogenates from mammals and insects. *Science* 155: 713.

Gentry, A.H. (1992). A synopsis of Bignoniaceae ethnobotany and economic botany. *Annals of the Missouri Botanical Garden* 79: 53–64.

Glenboski, L.L. (1983). *The ethnobotany of the Tukuna Indians, Amazónas, Colombia.* Biblioteca José Jerónimo Triana No. 4, Universidad Nacional de Colombia, Bogotá.

Good, K. (1995). Yanomami of Venezuela. Foragers or farmers – which came first? Pages 113–120 *in* L.E. Sponsel, org., *Indigenous peoples and the future of Amazonia. An ecological anthropology of an endangered world.* The University of Arizona Press, Tucson.

Grenand, P. (1980). *Introduction à l'étude de l'univers Wayãpi. Ethnoécologie des Indiens du Haut-Oyapock (Guyane Française).* Langues et Civilisations à Tradition Orale 40. SELAF, Paris.

Grenand, P. and Prevost, M-F. (1994). Les plantes colorantes utilisées en Guyane Française. *Journal d'Agriculture Tropicale et de Botanique Appliquée,* n.s. 36 (1): 139–172.

Grenand, P., Moretti, C. and Jacquemin, H. (1987). *Pharmacopées traditionelles en Guyane.* ORSTOM, Paris, France.

Hallé, F., Oldeman, R.A.A. and Tomlinson, P.B. (1978). *Tropical trees and forests. An architectural analysis.* Springer-Verlag, Berlin.

Hames, R. (1983a). The settlement pattern of a Yanomamö population block: a behavioural ecological interpretation. Pages 393–427 *in* R.B. Hames and W.T. Vickers, orgs, *Adaptive responses of native Amazonians.* Academic Press, New York.

Hames, R. (1983b). Monoculture, polyculture and polyvariety in tropical forest swidden cultivation. *Human Ecology* 11 (1): 13–34.

Harris, D.R. (1971). The ecology of swidden cultivation in the Upper Orinoco rain forest, Venezuela. *Geographical Review* 61: 475–495.

Harris, M. (1974). *Cows, pigs, wars and witches: the riddles of culture.* Fontana, Glasgow.

Hausen, B. (1981). *Woods injurious to human health: a manual.* Walter de Gruyter & Co., Berlin.

Heckel, E. (1897). *Les plantes médicinales et toxiques de la Guiane Française.* Protat Frères, Maçon, France.

Henderson, A. (1995). *The palms of the Amazon.* Oxford University Press.

Henderson, A., Galeano, G. and Bernal, R. (1995). *Field guide to the palms of the Americas.* Princeton University Press, New Jersey.

Holmes, R. (1995). Small is adaptive. Nutritional Anthropometry of native Amazonians. Pages 121–148 *in* L.E. Sponsel, org., *Indigenous peoples and the future of Amazonia. An ecological anthropology of an endangered world.* The University of Arizona Press, Tucson.

Huber, O. (1995). Vegetation. Pages 97–160 *in* J.A. Steyermark, P.E. Berry and B.K. Holst, eds, *Flora of the Venezuelan Guayana.* Volume 1. Introduction. Missouri Botanical Garden, St Louis.

Huber, O., Steyermark, J.A., Prance, G.T. and Alès, C. (1984). The vegetation of the Sierra Parima, Venezuela-Brazil: some results of recent exploration. *Brittonia* 36 (2): 104–139.

Im Thurn, E.F. (1883). *Among the Indians of Guiana.* Kegan Paul, London.

Kahn, F. (1997). *The palms of El Dorado.* ORSTOM – Editions Champflour, Marly-le-Roi.

Krukoff, B.A. and Barneby, R.C. (1970). Supplementary notes on American Menispermaceae. VI. *Memoirs of the New York Botanical Garden* 20 (2): 1–70.

Krukoff, B.A. and Smith, A.C. (1937). Rotenone-yielding plants of South America. *American Journal of Botany* 24: 573–587.

Kunstadter, P. (1979). Démographie. Pages 345–380 *in Ecosystèmes forestiers tropicaux.* UNESCO, Paris.

Kvist, L.P. and Holm-Nielsen, L.B. (1987). Ethnobotanical aspects of lowland Ecuador. *Opera Botanica* 92: 83–107.

La Rotta, C. (1983). *Observaciones etnobotánicas sobre algunas espécies utilizadas por la comunidad indígena Andoque (Amazonas Colombia).* DAINCO, Bogotá.

La Rotta, C. (1988). *Espécies utilizadas por la comunidad Miraña. Estudio etnobotánico.* FEN, Colombia.

Lachman-White, D.A., Adams, C.D. and Trotz, U. O.'D. (1987). *A guide to the medicinal plants of coastal Guyana.* Commonwealth Science Council, London.

Laudato, F., and Laudato, L. (1984). *Um mergulho na pré-história. Os últimos Yanomami?* Umberto Calderadora, Manaus.

Le Cointe, P. (1936). Les plantes à rotenone en Amazonie. *Revue de Botanique Appliquée* 16: 609–615.

Lewis, W.H., Elvin-Lewis, M. and Gnerre, M.C. (1987). Introduction to the ethnobotanical pharmacopeia of the Amazonian Jívaro of Peru. Pages

96–103 *in* A.J. Leeuwenberg, comp., *Medicinal and poisonous plants of the tropics*. Pudoc, Wageningen.

Lizot, J. (1972). Poisons yanõmamɨ de chasse, de guerre et de pêche. *Antropológica* 31: 3–20.

Lizot, J. (1974). Contribution a l'étude de la technologie yanõmamɨ. *Antropológica* 38: 15–33.

Lizot, J. (1978). L'économie primitive. *Libre* 4: 69–113.

Lizot, J. (1980). La agricultura yanõmamɨ. *Antropológica* 53: 3–93.

Lizot, J. (1984). *Les Yanõmamɨ centraux*. Cahiers de l'Homme, Éditions de L'EHESS, Paris.

Lizot, J. (1988). Los Yanõmamɨ. Pages 479–583 *in* J. Lizot, org., *Los aborígenas de Venezuela, vol. III, Etnologia contemporanea*. Fundación La Salle/Monte Avila, Caracas.

Lizot, J. (1996). *Introducción a la lengua yanomamɨ*. Morfologia. Vicariato Apostólico de Puerto Ayacucho, Caracas.

Lizot, J. (1997). L'exploitation des ressources naturelles chez les Yanõmamɨ: une stratégie globale. Pages 749–758, *in* C.M. Hladik, A. Hladik, H. Pagezy, O. Linares, G.J.A. Koppert and A. Froment, eds, *L'alimentation en forêt tropicale: interactions bioculturelles et perspectives de développement*. UNESCO, Paris.

Macmillan, G.J. (1995). *At the end of the rainbow? Gold, land and people in the Brazilian Amazon*. Earthscan, London.

Migliazza, E.C. (1972). *Yanomama grammar and intelligibility*. Doctoral thesis, Indiana University.

Migliazza, E.C. (1982). Linguistic prehistory and the refuge model in Amazonia. Pages 497–519 *in* G.T. Prance, org., *Biological diversification in the tropics*. Columbia University Press, New York.

Milliken, W. 1997. Traditional anti-malarial medicine in Roraima, Brazil. *Economic Botany* 51 (3): 212–237.

Milliken, W. and Albert, B. (1996). The use of medicinal plants by the Yanomami Indians of Brazil. *Economic Botany* 50 (1): 10–25.

Milliken, W. and Albert, B. (1997a). The use of medicinal plants by the Yanomami Indians of Brazil. Part II. *Economic Botany* 51 (3): 264–278.

Milliken, W. & Albert, B. (1997b). The construction of a new Yanomami round-house. *Journal of Ethnobiology* 17 (2): 215–233.

Milliken, W. and Ratter, J.A. (1989). *The vegetation of the Ilha de Maracá, Roraima, Brazil*. Royal Botanic Garden, Edinburgh.

Milliken, W., Miller, R.P., Pollard, S.R. and Wandelli, E.V. (1992). *Ethnobotany of the Waimiri Atroari Indians of Brazil*. Royal Botanic Gardens, Kew.

Morton, J.F. 1981. *Atlas of medicinal plants of Middle America*. C. Thomas, Springfield, Illinois.

Neel, J.V., Arends, T., Brewer, C., Chagnon, N., Gershowitz, H., Layrisse, M., Maccluer, J., Migliazza, E., Oliver, W., Salzano, F., Spielman, R., Ward, R. and Weitkamp, L. (1972). Studies on the Yanomama Indians. *Proceedings of the Fourth International Congress of Human Genetics. Amsterdam, Excerpta Medica*: 96–111.

Oliveira, J., Almeida, S.S., de Vilhena-Potyguara, R. and Lobato, J.C.B. (1991). Espécies vegetais produtoras de fibras utilizadas por comunidades Amazônicas. *Boletim do Museu Paraense Emílio Goeldi, Série Botânica* 7 (2): 393–428.

Parodi, J.L. (1988). The use of palms and other native plants in non-conventional, low cost rural housing in the Peruvian Amazon. *Advances in Economic Botany* 6: 119–129.

Pennington, T.D. (1997). *The genus Inga. Botany.* Royal Botanic Gardens, Kew.

Phillips, O. and Gentry, A.H. (1993 a). The useful plants of Tambopata, Peru: I. Statistical hypotheses tests with a new quantitative technique. *Economic Botany* 47 (1): 15–32.

Phillips, O. and Gentry, A.H. (1993 b). The Useful plants of Tambopata, Peru: II. Additional hypothesis testing in quantitative ethnobotany. *Economic Botany* 47 (1): 33–43.

Pinedo-Vasquez, M., Zarin, D., Jipp, P. and Chota-Inuma, J. (1990). Use-values of tree species in a communal forest reserve in northeast Peru. *Conservation Biology* 4: 405–416.

Plotkin, M.J. (1993). *Tales of a shaman's apprentice. An ethnobotanist searches for new medicines in the Amazon rain forest.* Penguin Books USA, New York.

Plowman, T.C., Leuchtmann, A. Blaney, C. and Clay, K. Significance of the fungus *Balansia cyperi* infecting medicinal species of *Cyperus* (Cyperaceae) from Amazonia. *Economic Botany* 44 (4): 452–462.

Prance, G.T. (1970). Notes on the use of plant hallucinogens in Amazonian Brazil. *Economic Botany* 24: 62–68.

Prance, G.T. (1972a). An ethnobotanical comparison of four tribes of Amazonian Indians. *Acta Amazonica* 2 (2): 7–27.

Prance, G.T. (1972b). Ethnobotanical notes from Amazonian Brazil. *Economic Botany* 26 (2): 221–237.

Prance, G.T. (1984). The use of edible fungi by Amazonian Indians. *Advances in Economic Botany* 1: 127–139.

Prance, G.T. (1990) The genus *Caryocar* (Caryocaraceae): an underexploited tropical resource. *Advances in Economic Botany* 8: 177–188.

Prance, G.T. and Silva, M.F. da (1973). Caryocaraceae. *Flora Neotropica Monograph* 12. Hafner Publishing Company, New York.

Prance, G.T., Balée, W., Boom, B.M. and Carneiro, R.L. (1987). Quantitative ethnobotany and the case for conservation in Amazonia. *Conservation Biology* 1 (4): 296–310.

RADAMBRASIL (1975). Folha NA.20 Boa Vista e parte das Folhas NA.21 Tumucumaque NB.20 Roraima e NB.21. *Levantamento de Recursos Naturais.* Vol. 8. DNPM, Rio de Janeiro.

Ramirez, H. (1994). *Le parler Yanomami des Xamatauteri.* Doctoral thesis, University of Aix en Provence.

Ramos, A.R. (1995). *Sanumá memories. Yanomami ethnography in times of crisis.* The University of Wisconsin Press, Madison.

Ramos, A.R., and Taylor, K.I. (eds.) (1979). *The Yanoama in Brazil.* IWGIA Document 37. IWGIA, Copenhagen.

Rizzini, C.T. (1971). *Árvores e madeiras úteis do Brasil.* Manual de dendrologia Brasileira. Editora Edgard Blücher Ltda, São Paulo.

Roosevelt, A.C., Costa, M.L. da, Machado, C.L., Michab, M., Mercier, N., Valladas, H., Feathers, J., Barnett, W., Silveirs, M.I. da, Henderson, A., Silva, J., Chernoff, B., Reese, D.S., Holman, J.A., Toth, N. and Schick, K. (1996). Paleoindian cave dwellers in the Amazon: the peopling of the Americas. *Science* 272: 373–384.

Roth, W.E. (1924). An introductory study of the arts, crafts, and customs of the Guiana Indians. *Annual Report of the Bureau of American Ethnology* 38: 25–745.

Schultes, R.E., and Holmstedt, B. (1968). The vegetal ingredients of the myristicaceous snuffs of the northwest Amazon. *Rhodora* 70: 113–160.

Schultes, R.E., and Raffauf, R.F. (1990). *The healing forest – medicinal and toxic plants of the northwest Amazonia.* Historical, ethno- and economic botany series, Vol. 2. Dioscorides Press, Portland, Oregon, USA.

Seitz, G.J. (1967). Epene, the intoxicating snuff powder of the Waika Indians and the Tucano medicine man, Agostino. Pages 315–338 *in* D.H. Efron, B. Holmstedt and N. Kline, eds, *Ethnopharmacologic search for psychoactive drugs.* U.S. Public Health Publication No. 1645.

Smith, N.J.H. (1980). Anthrosols and human carrying capacity in Amazonia. *Annals of the Association of American Geographers* 70 (4): 553–566.

Smole, W.J. (1976). *The Yanoama Indians. A cultural geography.* University of Texas Press, Austin.

Smole, W.J. (1989). Yanoama horticulture in the Parima Highlands of Venezuela and Brazil. *Advances in Economic Botany* 7: 115–128.

Smyth, A.J. (1975). Soils. Pages 34–47 *in* G.A.R. Wood, ed., *Cocoa.* Tropical Agriculture Series. Longman, London.

Spielman, R.S., Migliazza, E.C. Neel, J.V. Gershowitz, H. and Araúz, R.T. de (1979). The evolutionary relationships of two populations: a study of the Guaymí and the Yanomama. *Current Anthropology* 20 (2): 377–388.

Steinvorth de Goetz, I. (1969). *Uriji jami! Impresiones de viajes orinoquenses.* Asociación Cultural Humboldt, Caracas.

Steyermark, J.A., Berry, P.E. and Holst, B.K. (1995). *Flora of the Venezuelan Guayana.* Volume 1. Introduction. Missouri Botanical Garden, St Louis & Timber Press, Portland, Oregon.

Ter Welle, B.J.H. (1976). *Silica grains on woody plants of the Neotropics, especially Surinam.* Leiden Botanical Series 3: 107–142.

Valero, H. (1969). *Yanoáma: the story of a woman abducted by Brazilian Indians.* Allen & Unwin, London.

Valero, H. (1984). *Yo soy napë yoma. Relato de uma mujer raptada por los indígenas Yanomami.* Fundación La Salle de Ciencias Naturales, Caracas. Monografia 35.

Wilbert, J. and Simoneau, K. eds. (1990). *Folk literature of the Yanomami Indians.* UCLA Latin American Studies Vol. 73. University of California, Los Angeles.

Zerries, O. and Schuster, M. (1974). *Mahekodoteri.* Klaus Renner Verlag, Munich.

APPENDICES

Appendix I

Plant species discussed in this monograph

(with authors, families, Yanomami names, voucher specimen numbers, and locations in the text)

Note: This list does not cover all of the plant species collected during the present study. A large number of species which were collected solely for their medicinal properties have been omitted, in an effort to help to safeguard the intellectual property rights of the Yanomami (see 'Plants as medicine'). Species whose names are not given in bold type were not recorded among the Yanomami, and their inclusion in the text is purely for the purposes of comparison. In order to avoid confusion between dialects and orthographies, Yanomami names collected by other authors have not been listed here.

The majority of these data were collected during the present study. Where possible, voucher specimen numbers are cited as the primary source of reference. Collection series are abbreviated as follows: WM = this study; EF = Fuentes; JL = Lizot; GP = Prance. Other cited sources are as follows: AA = Anderson; CA = Alès; LC = Cocco and FI = Finkers. Where no number is given, n.c. indicates that the plant was identified during the present study but was not collected, and s.n. indicates the lack of a cited voucher specimen in a secondary source.

Locations of these species in the text are identified in the 'section' column as follows:

1 **Yanomami horticulture**
   a: Swidden clearings
   b: Ecological factors & swidden sites

2 **Plants as food**
   a: Cultivated food plants
   b: Wild food plants
   c: Fungi and flowers
   d: Food plant diversity
   e: Borderline species
   f: Plants for salt and water

3 **Drug and stimulant plants**
   a: Hallucinogens
   b: Tobacco and tobacco substitutes

4 **Plants employed in hunting and fishing**
   a: Weapons
   b: Poisoned arrows & interchangeable arrow-heads
   c: Dogs, men and plants

   d: Game ecology
   e: Insect ecology
   f: Fishing

5 **Plants for body adornment**
   a: Dress
   b: Body paints
   c: Scents and blossoms

6 **Plants for construction**
   a: Temporary shelters
   b: Structure and composition of the *yano*
   c: Changes in Yanomami construction
   d: Yanomami construction in the Amazonian context

7 **Plants for tools, implements & misc. uses**
   a: Fibres
   b: Basketry
   c: Other containers
   d: River transport

   e: Cutting tools
   f: Other uses

8 **Medicinal plants**
   a: The Yanomami pharmacopoeia
   b: Application and preparation techniques
   c: Evolution and efficacy of Yanomami medicinal plants
   d: Malaria, medicinal plants and the Yanomami: acquisition or experimentation?

9 **Plants for fire**

10 **Irritant plants**

11 **Plants in ritual, magic and myth**
   a: Shamanic spirits
   b: Magic, sorcery & ritual plants
   c: Plants in myth

Species	Family	Yanomami name	Source	Section
Abuta grandifolia (Mart.) Sandwith	Menispermaceae	*itahimosi hi*	WM1956	2b
Abuta grisebachii Triana & Planch.	Menispermaceae	*mãokori sina tʰotʰo*	WM2408	4b
Abuta imene (Mart.) Krukoff & Barneby	Menispermaceae	*mãokori sina tʰotʰo*	WM1786	4b
Abuta rufescens Aubl.	Menispermaceae	*werehe tʰotʰo*	WM2328	8a
Acacia polyphylla DC.	Leguminosae	*kayihi una tihi*	WM1836	1b, 4e
Aechmea sp.	Bromeliaceae	—	JF s.n.	2b
Alstroemeria sp.	Alstroemeriaceae	*manakaki*	WM2060	11b
Amasonia arborea Kunth	Verbenaceae	—	EF1044	5c
Amphirrhox surinamensis Eichl.	Violaceae	*maxopo mahi*	WM1708	6b
Amyris sp.	Rutaceae	—	EF0031	11b
Anacardium giganteum Hancock ex Engl.	Anacardiaceae	*oruxi hi*	WM1761	2a, 2b, 4d, 8a, 11a
Anacardium occidentale L.	Anacardiaceae	—	WM n.c.	2a
Anadenanthera peregrina (L.) Benth.	Leguminosae	*paara hi*	WM n.c.	3a
Ananas comosus (L.) Merr.	Bromeliaceae	—	WM n.c.	2a
Ananas sp.	Bromeliaceae	*yãma asiki*	WM2057	4a, 7a, 7e
Anaxagorea acuminata (Dun.) A.St.-Hil.	Annonaceae	*raina tihi*	WM1774	6b, 7a, 9
Andropogon bicornis L.	Gramineae	*pirima hiki*	WM1919	5a
Anemopaegma sp.	Bignoniaceae	—	—	6d
Aniba riparia (Mez) Kunth	Lauraceae	*tʰua mamo hi*	WM1993	6b, 6c
Annona amboiay Aubl.	Annonaceae	*totori mamo hi*	WM1793	3b
Anomospermum sp.	Menispermaceae	—	—	4b
Apeiba membranacea Spruce ex Benth.	Tiliaceae	*taitai ahi*	WM1847	1b, 7f
Apeiba? sp.	Tiliaceae	*warë unasi*	WM2069	7a
Aphelandra sp.	Acanthaceae	—	EF1018	3a
Aristolochia disticha Mast. vel aff.	Aristolochiaceae	*xuu tʰotʰo*	WM1713	8a, 8b
Aristolochia sp.	Aristolochiaceae	*ihuru tʰotʰo*	WM1976	11b
Arrabidaea sp.	Bignoniaceae	*matʰo tʰotʰo*	WM2018	6b, 11b
Aspidosperma nitidum Benth.	Apocynaceae	*hura sihi, poo hetʰo hi*	WM1734	7e, 8a, 8d
Aspidosperma sp.	Apocynaceae	*rahaka mahi*	WM1975	6b
Asplundia ponderosa R.E. Schult.	Cyclanthaceae	—	—	2f
Asplundia sp.	Cyclanthaceae	*yopo unaki*	WM2400	2f
Astrocaryum aculeatum Meyer	Palmae	*ëri si*	WM1821	2b, 8a, 9
Astrocaryum chambira Burret	Palmae	—	—	7f

Species	Family	Yanomami name	Source	Section
Astrocaryum gynacanthum Mart.	Palmae	*xoomo si*	WM2028	2b
Astrocaryum murumuru Mart.	Palmae	*maha si*	WM1883	2b, 2e, 4d
Astronium lecointei Ducke	Anacardiaceae	—	EF2194	2b
Bactris gasipaes Kunth	Palmae	*raxa si*	WM n.c.	2a, 4a, 4b, 7f, 11c
Bactris monticola Barb. Rodr.	Palmae	*mokamo si*	WM1983	2b, 3a, 6b
Bagassa guianensis Aubl.	Moraceae	*hapakara hi*	WM1763	2b, 2e, 4d, 7d, 9
Balansia cyperi Edg.	Clavicipitaceae	—	—	11b
Banisteriopsis lucida (Rich.) Small	Malpighiaceae	*xirixiri ãthe*	WM1859	4f
Bauhinia glabra Jacq.	Leguminosae	—	EF0024	5c
Bauhinia guianensis Aubl.	Leguminosae	*tiwakarama thotho*	WM1736	5c, 6a, 6b, 7a, 8a, 9, 11c
Begonia humilis Alt	Begoniaceae	*mor∙mor∙pë*	WM1837	1b
Bellucia grossularioides (L.) Triana	Melastomataceae	*pitima hi*	WM1922	2b, 4d
Bertholletia excelsa Humb. & Bonpl.	Lecythidaceae	*hawari hi*	WM n.c.	2b, 2d, 7c
Besleria laxiflora Benth.	Acanthaceae	*tixo nahe*	WM1813	3b
Bixa orellana L.	Bixaceae	*nara xihi*	WM n.c.	4a, 5a, 7a, 7f, 8b, 9, 11b, 11c
Bromelia tarapotina Ule	Bromeliaceae	—	EF0003	2b
Brosimum cf. *alicastrum* Swartz ssp. *bolivarense* (Pittier) C.C. Berg	Moraceae	—	EF0079	2b
Brosimum guianense (Aubl.) Huber	Moraceae	*ihuru makasi hi*	WM1778	2b, 2e, 4d, 9
Brosimum lactescens (S. Moore) C.C. Berg	Moraceae	*kãri nahi*	WM1762	4d, 9
Brosimum utile (Kunth) Pittier	Moraceae	*yaremaxi hi*	WM1996	7a, 11c
Buforrestia candolleana C.B. Clarke	Commelinaceae	—	WM1840	1b
Byrsonima aerugo Sagot	Malpighiaceae	*atama asi*	WM1995	2b
Caladium bicolor (Aiton) Vent.	Araceae	Various	WM1725	8a, 11b
Calathea altissima Horan	Marantaceae	—	JF s.n.	2a
Calathea cf. *majestica* (Linden) H.A. Kenn.	Marantaceae	*pisa asi*	WM2023	2b, 7b
Calathea aff. *mansonis* Koern.	Marantaceae	*makerema asi*	WM1748	1b, 2b
Calathea sp.	Marantaceae	*sarama asi*	WM1832	1b
Calathea sp.	Marantaceae	*kuma maki*	WM1913	2b
Callichlamys latifolia (Rich.) Schum.	Bignoniaceae	*matho thotho*	WM2019	6b, 11b
Canna indica L.	Cannaceae	*makerema asi*	WM2389	2a
Capsicum frutescens L.	Solanaceae	*prika aki*	WM n.c.	2a, 4f, 8a, 11c
Cardiospermum sp.	Sapindaceae	*õha moki*	WM n.c.	5a

Species	Family	Yanomami name	Source	Section
Carica papaya L.	Caricaceae	*rokoari si*	WM n.c.	2a
Caryocar glabrum (Aubl.) Pers.	Caryocaraceae	*xoxomo hi*	WM1798	2b, 11c
Caryocar pallidum A.C.Sm.	Caryocaraceae	*xoxomo hi*	JL2258	2b, 4f
Caryocar villosum (Aubl.) Pers.	Caryocaraceae	*ruapa hi*	WM1777	2b, 2d, 4d, 7f, 9
Casearia guianensis (Aubl.) Urban	Flacourtiaceae	*yãpi mamo hi*	WM2050	6b
Casearia javitensis Kunth	Flacourtiaceae	*waxia hi*	WM1787	6b
Cassia reticulata Willd.	Leguminosae	—	EF1191	11b
Catostema sp.	Bombacaceae	—	JL2240	2b
Cecropia aff. peltata L.	Moraceae	*tokori hanaki*	WM2345	8a
Cecropia sciadophylla Mart.	Moraceae	*kahu usihi*	WM2077	2b, 11c
Cecropia spp.	Moraceae	*xiki a*	WM2020/2022	4a, 4f, 5a, 7a, 7e
Cedrela sp.	Meliaceae	—	—	7f
Cedrelinga catenaeformis Ducke	Leguminosae	*apuru uhi*	WM1704	4f, 7c, 11a
Ceiba pentandra L.	Bombacaceae	*wari mahi*	WM n.c.	1b, 5c, 7c, 11a
Celosia argentea L.	Amaranthaceae	—	EF1080	5c
Centrolobium paraense Tul.	Leguminosae	*hewë nahi*	WM1989	6b, 6c, 9
Chimarrhis sp.	Rubiaceae	*xitokoma hi*	WM2390	—
Chrysochlamys weberbaueri Engl.	Guttiferae	*peima hi*	WM1723	5a
Chrysophyllum argenteum Jacq.	Sapotaceae	*naira hi*	WM2046	4d, 6b
Cissampelos pareira L.	Menispermaceae	—	JF s.n.	2b
Cissus erosa Rich.	Vitaceae	*momo tʰotʰo*	WM1901	10
Cissus setosa Roxb.	Vitaceae	—	—	10
Citrus aurantiifolia (Christm.) Swingle	Rutaceae	—	WM n.c.	2a
Citrus sinensis (L.) Osbeck	Rutaceae	—	WM n.c.	2a
Clarisia ilicifolia (Spreng.) Lanj. & Rossb.	Moraceae	*hihõ unahi*	WM1740	2b, 4d
Clarisia racemosa Ruiz & Pav.	Moraceae	*huyu hi*	WM n.c.	7d
Clathrotropis macrocarpa Ducke	Leguminosae	*wapo kohi*	WM1759	2b, 4b, 4c, 4d, 11c
Clavija lancifolia Desf.	Theophrastaceae	*horikima hi*	WM1953	2b
Clematis dioica L.	Ranunculaceae	*hemare motʰoki*	WM2436	8a
Clibadium sylvestre (Aubl.) Baill.	Compositae	*koaaxi hanaki*	WM1783	4f, 11b
Clusia spp.	Guttiferae	*pori pori tʰotʰo*	WM1982	7f, 8a, 11c
Cochlospermum orinocense (Kunth) Steud.	Cochlospermaceae	—	WM n.c.	7a
Cocos nucifera L.	Palmae	—	WM n.c.	2a

Species	Family	Yanomami name	Source	Section
Coix lacryma-jobi L.	Gramineae	—	WM n.c.	5a
Columnea picta Karst	Gesneriaceae	—	—	3b
Copaifera sp.	Leguminosae	*nara uhi*	WM n.c.	5b
Cordia cf. *lomatoloba* Johnst.	Boraginaceae	*hwaha nahi*	WM1973	9
Cordia nodosa Lam.	Boraginaceae	*xiho xihi*	WM2025	2b
Cordia tetrandra Aubl.	Boraginaceae	—	LC s.n.	7d
Corynostylis arborea (L.) Blake	Violaceae	—	EF1184	5b
Costus guanaiensis (Aubl.) Gaud.	Zingiberaceae	*naxuruma aki*	WM1700	8a
Costus spp.	Zingiberaceae	*naxuruma ahi*	WM1700/1710	1b
Couepia caryophylloides Ben.	Chrysobalanaceae	*wãro uhi*	WM2015	2b, 6b, 7c
Couepia sp.	Chrysobalanaceae	*hiha amo hi*	WM1885	2b
Couma macrocarpa Barb. Rodr.	Apocynaceae	*operema axi hi*	WM1882	2b, 4d, 5b, 7f
Couratari guianensis Aubl.	Lecythidaceae	*hotorea kosihi*	WM1912	2f, 4d, 4e, 7a
Coussapoa sp.	Moraceae	—	LC s.n.	4b
Crescentia cujete L.	Bignoniaceae	*maraka axihi*	WM n.c.	7a, 7c
Croton matourensis Aubl.	Euphorbiaceae	*ara usihi*	WM1923	6b, 7a
Croton palanostigma Klotzsch.	Euphorbiaceae	*kotoporo sihi*	WM2301	8a
Croton pullei Lanj.	Euphorbiaceae	—	GP s.n.	9
Curarea candicans (Rich.) Barneby & Krukoff	Menispermaceae	*mãoktori sina tʰotʰo*	WM2426	4b
Cyclanthus bipartitus Poit	Cyclanthaceae	—	—	2f
Cydista aequinoctialis (L.) Miers	Bignoniaceae	—	EF0018	5c, 7d
Cymbopogon citratus (Nees) Stapf.	Gramineae	*makiyuma hanaki,waihi hanaki, makiyuma xiki*	WM1775	8a
Cyperus articulatus L.	Cyperaceae	Various	WM1751	8a, 11b
Cyperus corymbosus Rottb.	Cyperaceae	—	—	11b
Dacryodes peruviana (Loes.) Lam.	Burseraceae	*mõra mahi*	WM2024	2b
Dialium guianense (Aubl.) Sandw.	Leguminosae	*paro koxiki*	WM2070	2b, 6c
Didymopanax morototoni (Aubl.) Decn. & Planch.	Araliaceae	—	EF1208	7d
Dieffenbachia bolivarana Bunting	Araceae	*xenoma a*	WM2376	8a
Dioclea aff. *malacocarpa* Ducke	Leguminosae	*kuapara tʰotʰo*	WM2351	2b
Dioscorea cf. *triphylla* Schimp.	Dioscoreaceae	—	EF1198	2b
Dioscorea piperifolia Humb. & Bonpl. ex Willd.	Dioscoreaceae	*kaxelema, yarimonahe*	WM1986	1b, 2b, 11b
Dioscorea trifida L.f.	Dioscoreaceae	*wãha aki, wãha urihitheri aki*	WM1849	2a, 2b, 11c

Species	Family	Yanomami name	Source	Section
Dioscorea sp.	Dioscoreaceae	*rai a*	WM n.c.	2b
Dipteryx odorata (Aubl.) Willd.	Leguminosae	*paetea sihi*	WM1958	11b
Distictella sp.	Bignoniaceae	—	—	6d
Dorstenia sp.	Moraceae	—	EF1052	11b
Dracontium asperum K. Koch vel aff.	Araceae	*kãtãrã ãsi*	WM1717	2b, 10
Drymonia coccinea Aubl.	Gesneriaceae	*hura si*	WM2325	8a
Duguetia lepidota (Miq.) Pulle	Annonaceae	*amaᵗʰa hi*	WM1803	3a, 6b
Duguetia spp.	Annonaceae	*hapoma hi*	WM n.c.	7f, 9
Duroia eriopila L.f.	Rubiaceae	*hera xihi*	WM1749	2b, 6b
Ecclinusa guianensis Eyma	Sapotaceae	*yãre hi*	WM2051	2b
Elizabetha leiogyne Ducke	Leguminosae	*ama hi*	WM1781	3a, 4d, 7a, 9
Elizabetha princeps Benth.	Leguminosae	—	EF s.n.	3a, 11c
Erechtites hieraciifolia Raf.	Compositae	*yavare nahasiki*	WM1744	3b
Eschweilera coriacea (A.DC.) Mori	Lecythidaceae	*hokoto uhi*	WM2007	4d, 6b, 7a, 7d
Eugenia flavescens DC. vel aff.	Myrtaceae	*pore hi*	WM2052	2b, 6b
Eugenia sp.	Myrtaceae	*korokoro sihi*	WM2065	6b
Euterpe precatoria Mart.	Palmae	*maima si*	WM1904	2b, 4d, 5b, 6c, 7b, 7f
Exellodendron barbatum (Hook.f.) Prance	Chrysobalanaceae	*ᵗʰomira asihi*	WM2006	7e
Favolus brasiliensis (Fr.) Fr.	Polyporaceae	*xokopë amoki*	WM1826	2b
Favolus spathulatus (Jungh.) Lév.	Polyporaceae	*uxirima amoki*	WM1827	2b
Ficus cf. *paraensis* (Miq.) Miq.	Moraceae	—	EF1108	11b
Filoboletus gracilis (Klotzsch ex Berk.) Singer	Tricholomataceae	*pihi wayorema amoki*	WM1823	2b
Fusaea longifolia (Aubl.) Saff.	Annonaceae	*hwapoma hi*	WM1881	2b, 3a, 6b, 7a
Geissospermum argenteum Woods.	Apocynaceae	—	JL2320	4b
Genipa americana L.	Rubiaceae	*hooma xihi*	WM2017	5b
Genipa spruceana Steyerm.	Rubiaceae	—	JL2262	5b
Geonoma baculifera (Poit.) Kunth	Palmae	*paa hanaki*	WM2034	5a, 5c, 6b, 6c, 11b
Geonoma deversa (Poit.) Kunth	Palmae	*warama si*	WM1729	2b, 2e, 6b, 9
Geonoma ? sp.	Palmae	*misikirima hanaki*	WM n.c.	5a
Geophila repens (L.) I.M. Johnst.	Rubiaceae	*mamo wai kiki, mamori kiki*	WM1712	8a
Gossypium barbadense L.	Malvaceae	*xinaru uhi*	WM1800	3b, 5a, 9
Gouania frangulaefolia (Schult.) Radlk.	Rhamnaceae	*wayawaya ᵗʰoᵗʰo*	WM1903	4e
Guadua spp.	Gramineae	*katana si*	WM1857/1930	4a, 4b, 7c, 7e, 7f, 10

Species	Family	Yanomami name	Source	Section
Guarea guidonia (L.) Sleum.	Meliaceae	**mãri hi**	WM1829	1b, 4d, 5c
Guatteria spp.	Annonaceae	**seisei unahi**	WM1951/2082	6b, 7a, 9, 11c
Gurania spinulosa (Poepp. & Endl.) Cogn.	Cucurbitaceae	**ihuru ᵗʰoᵗʰo**	WM1818	11b
Gynerium sagittatum (Aubl.) Beauv.	Gramineae	**xaraka a**	WM n.c.	3a, 4a, 8b, 9, 11c
Heliconia bihai (L.) L.	Heliconiaceae	**irohoma hanahi**	WM1767	1b, 2b, 2e, 7b, 8b, 11c
Helicostylis tomentosa (Poepp. & Endl.) Rusby	Moraceae	**xopa hi**	WM1758	2b, 2e, 4d, 9
Herrania lemniscata (Schomb.) R.E. Schult.	Sterculiaceae	**xahuturi unahi**	WM1796	2b, 9, 10, 11c
Heteropsis flexuosa (Kunth) Bunting	Araceae	**masi kiki**	WM2008	6b, 6c, 7a, 7b
Hieronima sp.	Euphorbiaceae	—	CA s.n.	5b
Himatanthus sp.	Apocynaceae	—	n.c.	5b
Hippeastrum puniceum (Lam.) Kuntze	Amaryllidaceae	**si waima a**	WM2342	5c, 8a
Hirtella sp.	Chrysobalanaceae	—	—	7e
Humiria balsamifera (Aubl.) A.St.-Hil.	Humiriaceae	**asoka hi**	WM1984	11b
Hymenaea courbaril L.	Leguminosae	**hatohato koxihi**	WM1877	2b
Hymenaea parvifolia Ducke	Leguminosae	**arõ kohi**	WM1735	2b, 4d, 5b, 8a, 11a
Inga acreana Harms	Leguminosae	**pahi hi**	WM2030	2b
Inga acuminata Benth.	Leguminosae	**ria moxiririma hi**	WM1865	2b, 8a, 8b, 11c
Inga alba (Sw.) Willm.	Leguminosae	**moxima hi**	WM1776	2b, 4a, 4d, 7c
Inga edulis Mart.	Leguminosae	**krepu uhi**	WM2081	1b, 2a, 2b
Inga illa T. Penn.	Leguminosae	—	—	2b
Inga myriantha Poepp. & Endl.	Leguminosae	—	EF1155	2b
Inga nobilis (Willd.) Benth.	Leguminosae	—	EF1094	2b
Inga paraensis Ducke vel aff.	Leguminosae	**toxa hi**	WM1811	2b, 4d
Inga pezizifera Benth.	Leguminosae	**kãi hi**	WM1911	2b
Inga pilosula (Rich.) Macbride	Leguminosae	**poataa hi**	WM1998	2b
Inga sarmentosa Glaz.	Leguminosae	**pooko hi**	WM1880	2b, 9
Inga scabriuscula Benth.	Leguminosae	—	JL2105	2b
Inga sp.	Leguminosae	**kãi hi**	WM n.c.	2b
Ipomoea batatas (L.) Lam.	Convolvulaceae	**hõkõmo ᵗʰoᵗʰo**	WM n.c.	2a
Ipomoea cf. *batatas* (L.) Lam.	Convolvulaceae	**hõkõmo urihiᵗʰeri a**	WM n.c.	2b
Iriartea deltoidea Ruiz & Pav	Palmae	**kripi si**	WM n.c.	4a, 7c
Iriartella setigera (Mart.) Wendl.	Palmae	**horoma si**	WM2009	3a, 4b, 10, 11c
Iryanthera juruensis Warb.	Myristicaceae	**sikãri a**	WM2049	4d, 6b

Species	Family	Yanomami name	Source	Section
Iryanthera laevis Mgf	Myristicaceae	*sikāri a*	WM2048	6b
Iryanthera ulei Warb.	Myristicaceae	—	EF1222	2b
Ischnosiphon arouma (Aubl.) Koern.	Marantaceae	*mokuruma si*	WM2087	1b, 3a, 7b
Ischnosiphon obliquus (Rudge) Koern.	Marantaceae	*mokuruma si*	BA s.n.	3a, 7b
Isertia hypoleuca Benth.	Rubiaceae	*yāpi uhi*	WM1875	2b
Isertia parviflora Vahl	Rubiaceae	—	EF1018	3a
Jacaranda copaia (Aubl.) D. Don	Bignoniaceae	*xitopari hi*	WM1732	1b, 7f
Jacaratia digitata (Poepp. & Endl.) Solms	Caricaceae	*rihuwari si*	WM1839	1b, 2b, 4e
Jessenia bataua (Mart.) Burret	Palmae	*koanari si*	WM n.c.	2b, 4a, 4b, 5b, 6b, 7b, 7f
Jessenia polycarpa Karst.	Palmae	—	JL2316	2b
Justicia pectoralis Jacq.	Acanthaceae	Various	WM2055	5c, 11b, 11c
Justicia pectoralis Jacq. var. *stenophylla* Leonard	Acanthaceae	*maxahara hanaki*	WM1784	3a
Lacunaria jenmani (Oliv.) Ducke	Quinaceae	*paari makasi hi*	WM2003	2b
Lagenaria siceraria (Molina) Standl.	Cucurbitaceae	*horohoto tʰotʰo*	WM1921	7c
Lantana trifolia L.	Verbenaceae	—	JF s.n.	5c
Lentinus tephroleucus Mont.	Polyporaceae	*haya kasiki*	WM1825	2b
Licania heteromorpha Benth	Chrysobalanaceae	*maraka axihi*	WM2040	4d, 6b, 7c
Licania kunthiana Hook.f.	Chrysobalanaceae	*mraka nahi*	WM2035	5c, 6b
Licania macrophylla Benth.	Chrysobalanaceae	—	—	4a
Licania cf. *polita* Spruce ex Hook.f.	Chrysobalanaceae	*xihini hi*	WM2045	6b
Licaria aurea (Huber) Kosterm.	Lauraceae	*hokōma hi*	WM2002	3b, 6b, 6c, 7c
Lonchocarpus cf. *chrysophyllus* Kleinh.	Leguminosae	*xina āthe*	WM1963	4f, 7f, 10
Lonchocarpus utilis A.C.Sm.	Leguminosae	*ketʰā āthe*	WM1967	4f, 10
Mabea sp.	Euphorbiaceae	—	n.c.	4a
Machaerium macrophyllum Benth.	Leguminosae	*akanasima tʰotʰo*	WM2013	7a, 11b
Machaerium quinata (Aubl.) Sandw. vel aff.	Leguminosae	*rāasirima tʰotʰo*	WM2388	8a
Malpighia punicifolia L.	Malpighiaceae	—	WM n.c.	2a
Mangifera indica L.	Anacardiaceae	—	WM n.c.	2a
Manihot esculenta Crantz	Euphorbiaceae	*hutu si*	WM n.c.	2a
Manilkara bidentata (A.DC.) Chev.	Sapotaceae	*nai hi*	2b	2b, 4d, 6b, 6c
Manilkara huberi (Ducke) Standl.	Sapotaceae	*xaraka ahi*	WM2042	6b, 6c, 11b
Maprounea guianensis Aubl.	Euphorbiaceae	*yipi hi*	WM2072	2b
Maquira calophylla (Poepp. & Endl.) C.C. Berg	Moraceae	*teria hi*	WM2029	2b

Species	Family	Yanomami name	Source	Section
Maranta arundinacea L.	Marantaceae	*hore kiki*	WM1726	8a, 11b
Marasmius cubensis (Berk. & Curtis) Sacc.	Tricholomataceae	*moka uku*	WM1824	2b
Margaritaria nobilis (L.) Müll. Arg.	Euphorbiaceae	*yãpi mamohi*	WM1838	1b, 4d
Martinella sp.	Bignoniaceae	—	—	6d
Martiodendron sp.	Leguminosae	*rapa hi*	WM2076	6c
Martiodendron sp.	Leguminosae	*paxo hi*	WM2054	6b
Mauritia aculeata Kunth	Palmae	—	AA s.n.	2b
Mauritia flexuosa L.	Palmae	*rioko si*	WM n.c.	2b, 4e, 7b, 7f
Mauritiella armata (Mart.) Burret	Palmae	*kuai si*	WM n.c.	2b
Maximiliana maripa (Corrêa) Drude	Palmae	*ökorasi si*	WM n.c.	2b, 3a, 4e, 6b, 7a, 7c, 7f
Melloa sp.	Bignoniaceae	—	—	6d
Mesechites trifida (Jacq.) Müll. Arg.	Apocynaceae	—	JL2094	11b
Miconia lateriflora Cogn.	Melastomataceae	*xama siosiki*	WM1771	3b
Mierandra rossiana R.E. Schult.	Euphorbiaceae	*momo hi*	JL2176	2b, 11c
Micropholis melinoniana Pierre	Sapotaceae	*apia hi*	WM1915	2b, 4d
Minquartia guianensis Aubl.	Olacaceae	—	—	6d
Monotagma sp.	Marantaceae	*kopari hanaki*	WM n.c.	2b, 2e
Monstera adansonii Schott	Araceae	*xãã a*	WM2432	8a
Moronobea sp.	Guttiferae	*mai kohi*	WM1876	4a, 7e
Mouriri myrtifolia Spruce	Melastomataceae	—	JL2051	4a
Mouriri nervosa Pilg.	Melastomataceae	*tihi hi*	WM2384	7f
Mouriri sagotiana Triana	Melastomataceae	—	JL2158	4a
Mucuna urens (L.) Medik.	Leguminosae	*wakõwakõ axi ħoħo*	WM1769	10
Musa spp.	Musaceae	*koraha si*	WM n.c.	2a
Mussatia sp.	Bignoniaceae	—	—	6d
Myrcia sp.	Myrtaceae	*totori mamo hi*	WM2033	2b, 6b
Myroxylon balsamum (L.) Harms	Leguminosae	—	EF1057	5a, 11b
Nectandra sp.	Lauraceae	*rapa mahi*	WM2063	6b
Nicotiana tabacum L.	Solanaceae	*pee nahe*	WM n.c.	3b, 8a
Ocotea sp.	Lauraceae	*iroma sihi*	WM2073	6c
Oenocarpus bacaba Mart.	Palmae	*hoko si*	WM n.c.	2b, 4a, 4e, 7f, 11c
Olyra latifolia L.	Gramineae	*purunama usi*	WM2001	7f
Olyra sp.	Gramineae	*purunama asi*	WM1834	1b, 3a

Species	Family	Yanomami name	Source	Section
Orbignya spectabilis (Mart.) Burret	Palmae	—	AA s.n.	2b
Ormosia sp.	Leguminosae	—	n.c.	11c
Orthoclada laxa (Rich.) Beauv.	Gramineae	*tʰomi koxi*	WM1835	1b
Paragonia pyramidata (Rich.) Bur.	Bignoniaceae	—	EF0034	4b
Parinari sp.	Chrysobalanaceae	—	—	7e
Passiflora coccinea Aubl.	Passifloraceae	*naxuruma tʰotʰo*	WM2032	2b
Passiflora fuchsiiflora Hemsley	Passifloraceae	*ihuru tʰotʰo*	WM1801	11b
Passiflora longiracemosa Ducke	Passifloraceae	—	EF0005	5c
Passiflora vitifolia Kunth	Passifloraceae	—	JF s.n.	2b
Paullinia cf. *pinnata* L.	Sapindaceae	—	EF0011	2b
Paullinia? sp.	Sapindaceae	—	EF0148	3a
Peperomia magnoliifolia (Jacq.) A. Dietr.	Piperaceae	*në wãri hanaki*	WM1899	8a
Peperomia rotundifolia (L.) Kunth	Piperaceae	*oru kiki wite, hura si*	WM1720	8a
Perebea angustifolia (Poepp. & Endl.) C.C.Berg	Moraceae	*rëxë hi*	WM2062	2b
Perebea guianensis Aubl.	Moraceae	*peesisima hi*	WM1898	2b
Persea americana Mill.	Lauraceae	*ahuari hi*	WM n.c.	2a
Pharus virescens Doell.	Gramineae	*xikirima hanaki*	WM1845	1b
Phenakospermum guyannense (L.C. Rich.) Endl. ex Miq.	Musaceae	*ruru asi*	WM1906	1b, 2a, 2b, 2d,6a, 6b, 7b, 11c
Philodendron cf. *divaricatum* K. Krause	Araceae	*morokoma tʰotʰo*	WM2071	7b
Philodendron hylaeae G.S. Bunting	Araceae	*ripo tʰotʰo*	WM1731	11b
Philodendron solimoesensis A.C. Sm.	Araceae	*puu tʰotʰo*	WM1790	8a
Phlebodium decumanum Willd.	Polypodiaceae	*tokosi hanaki*	WM1890	8a
Phyllanthus brasiliensis (Aubl.) Müll. Arg.	Euphorbiaceae	*parapara hi, oko hi*	WM1782	4f
Phytolacca ririnoides Kunth & Bouché	Phytolaccaceae	*kripiari hi*	WM1928	8a
Picramnia macrostachya Klotzsch ex Engl.	Simaroubaceae	—	JL2106	5b
Picramnia spruceana Engl.	Simaroubaceae	*koe axihi*	WM1739	4a, 5b, 7b, 7c, 8a, 11b
Pinzona coriacea Mart. & Zucc.	Dilleniaceae	*maani tʰotʰo*	WM1795	2f, 10
Piper arborea Aubl.	Piperaceae	*kahu mahi*	WM1804	8a
Piper bartlingianum (Miq.) C.DC.	Piperaceae	*paari maheto siki*	WM1703	3b
Piper demeraranum (Miq.) C.DC.	Piperaceae	*paari mahekoki*	WM1788	3b
Piper dilatatum Rich.	Piperaceae	—	EF0032	4b
Piper francovilleanum C.DC.	Piperaceae	*mosipoima hanaki*	WM1924	3b

143

Species	Family	Yanomami name	Source	Section
Piper interitum Trel. & Yunck.	Piperaceae	—	—	3b
Piper piscatorium Trel. & Yunck.	Piperaceae	—	—	3b
Piper sp.	Piperaceae	*mahekoma hi*	WM1848	1b
Piptadenia sp.	Leguminosae	*ërama kiki*	WM1860	1b
Pleurotus flabellatus (Berk. & Br.) Sacc.	Polyporaceae	*pokara amoki*	WM1851	2b
Plukenetia abutaefolia (Ducke) Pax & Hoffm.	Euphorbiaceae	*xörahe ʰoʰo*	WM2068	2b
Pogonophora schomburgkiana Miers	Euphorbiaceae	*tihitihi nahi*	WM2043	6b
Polyporus grammocephalus Berk.	Polyporaceae	*ara amoki*	WM1822	2b
Posadaea sphaerocarpa Cogn.	Cucurbitaceae	*pora axi*	WM1868	2b, 7c, 11b
Posqueria latifolia (Rudge) Roem. & Schult.	Rubiaceae	*weri nahi*	WM2026	5c
Pothomorphe peltata (L.) Miq.	Piperaceae	*mahekoma hanaki*	WM2346	1b, 8a
Pourouma bicolor Mart. ssp. *digitata* (Tréc.) C.C. Berg & van Heusden	Moraceae	*öema ahi*	WM1768	2b, 4a, 4d, 7f, 8c, 9
Pourouma melinonii Benoist	Moraceae	*waraka ahi*	WM1910	2b
Pourouma minor Benoist	Moraceae	*warihinama usihi*	WM1753	4d, 9
Pourouma ovata Tréc.	Moraceae	*mominari usihi*	WM2041	4d, 6b
Pourouma tomentosa Miq. ssp. *persecta* Standl. ex C.C. Berg	Moraceae	*kahu akahi*	WM2044	6b
Pouteria caimito (Ruiz & Pav.) Radlk.	Sapotaceae	*paxo hwätemo hi*	WM2010	2b, 6b
Pouteria eladantha Sandw.	Sapotaceae	*hörömo nahi*	WM2039	2b, 6b
Pouteria hispida Eyma	Sapotaceae	*yäwa xihi*	WM2064	2b, 6b
Pouteria venosa (Mart.) Baehni ssp. *amazonica* Penn.	Sapotaceae	*mäiko nahi*	WM2074	6b, 6c
Pouteria sp.	Sapotaceae	*poxe mamokasi hi*	WM2084	6c
Pradosia surinamensis (Eyma) T.D. Penn.	Sapotaceae	*werihisi hi*	WM2083	2b
Protium fimbriatum Swart	Burseraceae	*weyeri hi, mari hi*	WM1765	4e, 6b
Protium polybotryum (Turcz.) Engl.	Burseraceae	*hwaximo kohosi hi*	WM1991	2b
Protium spruceanum (Benth.) Engl.	Burseraceae	*warapa kohi*	WM2370	8a
Protium spp.	Burseraceae	*warapa kohi*	WM n.c.	5b, 8b, 9
Psammisia guianensis Klotzsch	Ericaceae	*teosi hanaki*	WM2335	—
Pseudolmedia laevigata Tréc.	Moraceae	*hayi hi*	WM1806	2b, 4d, 9, 11c
Pseudolmedia laevis (Ruiz & Pav.) Macbr.	Moraceae	*aso asihi*	WM1741	2b, 4d, 6b, 9, 11c
Psidium guajava L.	Myrtaceae	—	WM n.c.	2a
Psidium sp.	Myrtaceae	—	EF0027	11b

Species	Family	Yanomami name	Source	Section
Psychotria poeppigiana Müll. Arg.	Rubiaceae	—	EF1076	11c
Psychotria ulviformis Steyerm.	Rubiaceae	*ahete hanaki*	WM1705	11b
Quiina florida Tul. vel aff.	Quinaceae	*naxuruma ahi*	WM2037	2b, 6b, 6c
Randia sp.	Rubiaceae	—	JL s.n.	5c
Renealmia alpinia (Aubl.) Maas	Zingiberaceae	*mãokori sinaki*	WM1971	5a, 8b
Renealmia floribunda K. Schum.	Zingiberaceae	*nini kiki, hayeyama si, Omoari si, haro hanaki*	WM1969	8a
Rheedia benthamiana Triana & Planch.	Guttiferae	*oruhe hi*	WM2085	2b
Rheedia macrophylla (Mart.) Triana & Planch.	Guttiferae	*kotaki axihi*	WM1853	1b, 2b
Rhodostemonodaphne grandis (Mez) Rohwer	Lauraceae	*ruru hi*	WM1999	7c
Rinorea lindeniana (Tul.) Kuntze	Violaceae	*okoraxi hi*	WM1895	4a, 6b, 7e, 9
Rinorea riana (DC.) Kuntze	Violaceae	—	EF1122	5a
Saccharum officinarum L.	Gramineae	*puu si*	WM n.c.	2a
Sagotia racemosa Baill.	Euphorbiaceae	*sina hi*	WM1805	4d, 9
Sanchezia sp.	Acanthaceae	*ata hi*	WM2363	2f, 5c
Scheelea martiana Burret	Palmae	—	AA s.n.	2b
Sciaphila purpurea Benth.	Triuridaceae	*manakaki huëri*	WM1743	11b
Sclerolobium sp.	Leguminosae	*paya hi*	WM2075	6c
Securidaca diversifolia (L.) Blake	Polygalaceae	*kumi tʰotʰo*	WM1974	11a, 11b
Serjania grandifolia Sagot.	Sapindaceae	*xokopë äthe*	WM1772	4f
Siparuna decipiens (Tul.) DC.	Monimiaceae	*maharema ahi*	WM1724	6b
Siparuna guianensis Aubl.	Monimiaceae	*mõe hi*	WM1894	6b, 8a, 8b
Sloanea macrophylla Benth. ex Turcz. vel aff.	Elaeocarpaceae	*akapa ahi*	WM2038	6b
Socratea exorrhiza (Mart.) H. Wendl.	Palmae	*manaka si*	WM1866	2b, 2e, 4a, 4b, 6b, 7c, 7f, 11c
Solandra grandiflora Sw.	Solanaceae	*tʰooro a*	WM1766	5c
Solanum asperum Rich.	Solanaceae	*hwãha xihi*	WM1754	10
Solanum oocarpum Sendtn.	Solanaceae	*pokara mamo kasiki*	WM1952	11b
Sorocea muriculata Miq. ssp. *uaupensis* C.C. Berg	Moraceae	*yĩpi hi*	WM2349	2b, 11b
Spathiphyllum sp.	Araceae	—	EF1072	5c
Spondias mombin L.	Anacardiaceae	*pirima ahi tʰotʰo*	WM1733	1b, 2b, 2e, 4d, 8a, 9
Sterculia pruriens (Aubl.) K. Schum.	Sterculiaceae	*raxana tihi*	WM2012	7a
Strychnos guianensis Aubl.	Loganiaceae	*mãokori tʰoxi*	WM2399	4b

Species	Family	Yanomami name	Source	Section
Strychnos cf. hirsuta Spruce ex Benth.	Loganiaceae	māokori	WM2423	4b
Stryphnodendron pulcherrimum (Willd.) Hochr.	Leguminosae	maxahara sihi	WM1997	7a
Swartzia schomburgkii Benth.	Leguminosae	xitokoma hi	WM1926	8a
Swartzia sp.	Leguminosae	paira ahi	WM1909	4a, 4e
Symphonia globulifera L.f.	Guttiferae	mai kohi	WM2380	4a, 7e
Syngonium vellozianum Schott	Araceae	poroma tʰotʰo	WM1756	10, 11b
Tabebuia capitata (Bur. & K. Schum.). Sandw.	Bignoniaceae	masihanari kohi	WM2053	6b, 11a, 11c
Tabebuia guayacan (Seem.) Hemsl.	Bignoniaceae	—	JL2278	7d
Tabernaemontana angulata Mart. ex Müll. Arg.	Apocynaceae	akä hi	WM1925	4f
Tabernaemontana heterophylla Vahl	Apocynaceae	—	JL s.n.	5b
Tabernaemontana macrocalyx Müll. Arg.	Apocynaceae	asokoma hi	WM2425	8a
Tabernaemontana muelleriana Mart. ex Müll. Arg	Apocynaceae	—	—	4f
Tabernaemontana sananho Ruiz & Pav.	Apocynaceae	tʰoru hwätemo hi	WM1820	2b, 4b, 4f, 8a
Tachigali aff. myrmecophila (Ducke) Ducke	Leguminosae	kätäena hi	WM2047	6b, 6d, 8b
Tachigali paniculata Aubl.	Leguminosae	—	JL2102	4c, 11b
Talisia cf. pedicellaris (Sagot) Radlk.	Sapindaceae	mako ahi	WM2031	2b
Tanaecium nocturnum (Barb. Rodr.) Bur. ex K. Schum.	Bignoniaceae	puu tʰotʰo moki, paraparamo natʰoki	WM1719	8a
Tetragastris altissima (Aubl.) Swart	Burseraceae	xoko hwaka hi	WM2027	2b
Theobroma bicolor Humb. & Bonpl.	Sterculiaceae	himara amohi	WM1760	1b, 2b, 2e, 7a
Theobroma cacao L.	Sterculiaceae	poroa unahi	WM1816	1b, 2b, 9, 11c
Theobroma microcarpum Mart.	Sterculiaceae	prōö masihi	WM1858	2b
Theobroma subincanum Mart.	Sterculiaceae	waiporo unahi	WM1920	2b, 4d, 9
Thoracocarpus bissectus (Vell.) Harl.	Cyclanthaceae	yäri tʰotʰo	WM1994	6b, 6d
Tocoyena sp.	Rubiaceae	—	JL s.n.	5c
Trichilia pleeana (A. Juss.) DC.	Meliaceae	hwatʰo ahi	WM1831	1b, 4d
Trichilia sp.	Meliaceae	akanaxi ahi	WM2011	1b, 6b
Tynanthus polyanthus (Bur.) Sandwith	Bignoniaceae	—	EF1024	7d
Uncaria guianensis (Aubl.) Gmel.	Rubiaceae	ërama tʰotʰo	WM1738	4f, 8a, 8b, 11b
Urera baccifera (L.) Gaud.	Urticaceae	ira naki	WM1722	1b, 7f, 8a
Urera caracasana (Jacq.) Griseb.	Urticaceae	apinaki	WM1972	1b, 3a, 7f
Vataireopsis cf. surinamensis Lima	Leguminosae	yoroko axihi	WM2080	10, 11a
Virola elongata (Benth.) Warb.	Myristicaceae	yãhoana hi	WM1992	3a, 4b, 6b, 9, 11a

Species	Family	Yanomami name	Source	Section
Virola sebifera Aubl.	Myristicaceae	—	EF1004	3a
Virola theiodora (Benth.) Warb.	Myristicaceae	—	GP9638	3a, 11c
Vismia angusta Miq.	Guttiferae	*yoasi hi*	WM1856	8a, 8c
Vismia cayennensis (Jacq.) Pers.	Guttiferae	*yoasi hi, witari mahi*	WM1867	8c
Vismia guianensis (Aubl.) Choisy	Guttiferae	*siiriama sihi*	WM2360	4e, 8a
Wulffia baccata Kuntze	Compositae		JF s.n.	2b
Xanthosoma sagittifolium (L.) Schott	Araceae	*poroma hanaki*	WM1884	10, 11b
Xanthosoma sp.	Araceae	*aria kiki*	WM n.c.	2a
Xylopia sp.	Annonaceae	*yao nahi*	WM1979	6b
Zanthoxylum pentandrum (Aubl.) R. Howard	Rutaceae	*naharä hi*	WM1979	8a
Zanthoxylum rhoifolium Lam.	Rutaceae	*mano wai hanaki, nahiri hanaki*	WM2327	8a
Zea mays L.	Gramineae	*yonomo si*	WM n.c.	2a
Zingiber officinale Roscoe	Zingiberaceae	*amatha kiki*	WM1931	3b, 4c, 8a, 8c
Zizyphus cinnamomum Triana & Planch.	Rhamnaceae	*mirama asihi*	WM1871	2b
Zollernia paraensis Huber	Leguminosae	*opo sihi, uki sihi*	WM1959	3b, 6b
Indet.	Bombacaceae	*makuta asihi*	n.c.	5c
Indet.	Gramineae	*höröma siki*	n.c.	7b
Indet.	Lauraceae	*pohara mamohi*	WM1843	1b
Indet.	Lauraceae	*warë amohi*	WM1862	9
Indet.	Palmae	*misikirima hanaki*	n.c.	5a
Indet.	—	*wayapapë*	WM1833	1b
Indet.	—	*oko hi*	WM1846	1b

Appendix II

Yanomami names of plant species discussed in the text

ahüari hi	*Persea americana*	*hore kiki*	*Maranta arundinacea*
ahete hanaki	*Psychotria ulviformis*	*horikima hi*	*Clavija lancifolia*
akanasima tʰotʰo	*Machaerium macrophyllum*	*horokoto tʰotʰo*	*Lagenaria siceraria*
		horoma si	*Iriartella setigera*
akanaxi ahi	*Trichilia* sp.	*hõrõmo nahi*	*Pouteria cladantha*
akapa ahi	*Sloanea macrophylla* vel aff.	*hotorea kosihi*	*Couratari guianensis*
		hura si	*Drymonia coccinea*
akiã hi	*Tabernaemontana angulata*	*hura si*	*Peperomia rotundifolia*
		hura sihi	*Aspidosperma nitidum*
ama hi	*Elizabetha leiogyne*	*hutu si*	*Manihot esculenta*
amatʰa hi	*Duguetia lepidota*	*huyu hi*	*Clarisia racemosa*
amatʰa kiki	*Zingiber officinale*	*hwaha nahi*	*Cordia* cf. *lomatoloba*
amixi hanaki	*Justicia pectoralis*	*hwãha xihi*	*Solanum asperum*
apia hi	*Micropholis melinoniana*	*hwapoma hi*	*Fusaea longifolia*
apinaki	*Urera caracasana*	*hwatʰo ahi*	*Trichilia pleeana*
apina siki	*Urera* sp(p).	*hwaximo kohosi hi*	*Protium polybotryum*
apüru uhi	*Cedrelinga catenaeformis*	*hwëri kiki yai*	*Cyperus articulatus*
ara amoki	*Polyporus grammocephalus*	*ihuru makasi hi*	*Brosimum guianense*
		ihuru tʰotʰo	*Aristolochia* sp.
ara usihi	*Croton matourensis*	*ihuru tʰotʰo*	*Gurania spinulosa*
aria kiki	*Xanthosoma* sp.	*ihuru tʰotʰo*	*Passiflora fuchsiiflora*
arõ kohi	*Hymenaea parvifolia*	*ira naki*	*Urera baccifera*
aroari kiki	*Cyperus articulatus*	*irokoma hanaki*	*Heliconia bihai*
aso asihi	*Pseudolmedia laevis*	*iroma sihi*	*Ocotea* sp.
asoka hi	*Humiria balsamifera*	*itahimosi hi*	*Abuta grandifolia*
asokoma hi	*Tabernaemontana macrocalyx*	*këtʰã ãtʰe*	*Lonchocarpus utilis*
		kahu akahi	*Pourouma tomentosa* ssp. *persecta*
ata hi	*Sanchezia* sp.		
atama asi	*Byrsonima aerugo*	*kahu mahi*	*Piper arboreo*
ërama kiki	*Piptadenia* sp.	*kahu usihi*	*Cecropia sciadophylla*
ërama tʰotʰo	*Uncaria guianensis*	*kãi hi*	*Inga* sp(p).
ëri si	*Astrocaryum aculeatum*	*kãri nahi*	*Brosimum lactescens*
hapakara hi	*Bagassa guianensis*	*kãtãena hi*	*Tachigali* aff. *myrmecophila*
hapoma hi	*Duguetia* spp.		
haro hanaki	*Renealmia floribunda*	*katana si*	*Guadua* spp.
hatohato koxihi	*Hymenaea courbaril*	*kãtãrã ãsi*	*Dracontium asperum* vel aff.
hawari hi	*Bertholletia excelsa*		
haya kasiki	*Lentinus tephroleucus*	*kaxetema*	*Dioscorea piperifolia*
hayayama si	*Renealmia floribunda*	*kayihi una tihi*	*Acacia polyphylla.*
hayi hi	*Pseudolmedia laevigata*	*koaaxi hanaki*	*Clibadium sylvestre*
hemare motʰoki	*Clematis dioica*	*koamaxi kiki*	*Cyperus articulatus*
hera xihi	*Duroia eriopila*	*koanari si*	*Jessenia bataua*
hewë nahi	*Centrolobium paraense*	*koe axihi*	*Picramnia spruceana*
hiha amo hi	*Couepia* sp.	*kopari hanaki*	*Monotagma* sp.
hihõ unahi	*Clarisia ilicifolia*	*koraha si*	*Musa* spp.
himara amohi	*Theobroma bicolor*	*korokoro sihi*	*Eugenia* sp.
hoko si	*Oenocarpus bacaba*	*kotaki axihi*	*Rheedia macrophylla*
hõkõma hi	*Licaria aurea*	*kotoporo sihi*	*Croton palanostigma*
hõkõmo tʰotʰo	*Ipomoea batatas*	*krepu uhi*	*Inga edulis*
hõkõmo urihitʰeri a	*Ipomoea* cf. *batatas*	*kripi si*	*Iriartea deltoidea(?)*
hokoto uhi	*Eschweilera coriacea*	*kripiari hi*	*Phytolacca rivinoides*
hooma xihi	*Genipa americana*	*kuai si*	*Mauritiella armata*

kuapara t^hot^ho — *Dioclea* aff. *malacocarpa*
kuma maki — *Calathea* sp.
kumi t^hot^ho — *Securidaca diversifolia*
maani t^hot^ho — *Pinzona coriacea*
maha si — *Astrocaryum murumuru*
maharema ahi — *Siparuna decipiens*
mahekoma hanaki — *Pothomorphe peltata*
mahekoma hi — *Piper* sp.
mai kohi — *Moronobea* sp.
mai kohi — *Symphonia globulifera*
māiko nahi — *Pouteria venosa* ssp. *amazonica*
maima si — *Euterpe precatoria*
makerema asi — *Calathea* aff. *mansonis**
makerema asi — *Canna indica**
makiyuma hanaki — *Cymbopogon citratus*
makiyuma xiki — *Cymbopogon citratus*
mako ahi — *Talisia* cf. *pedicellaris*
mamo wai hanaki — *Zanthoxylum rhoifolium*
mamo wai kiki — *Geophila repens*
mamori kiki — *Geophila repens*
manaka si — *Socratea exorrhiza*
manakaki — *Alstroemeria* sp.
manakaki hwëri — *Sciaphila purpurea*
mani hi — *Protium fimbriatum*
māokori — *Strychnos* cf. *hirsuta*
māokori sina t^hot^ho — *Abuta grisebachii*
māokori sina t^hot^ho — *Abuta imene*
māokori sina t^hot^ho — *Curarea candicans*
māokori sinaki — *Renealmia alpinia*
māokori t^hoxi — *Strychnos guianensis*
maraka axihi — *Crescentia cujete*
maraka axihi — *Licania heteromorpha*
māri hi — *Guarea guidonia*
marixi uki — *Cyperus articulatus*
masi kiki — *Heteropsis flexuosa*
masihanari kohi — *Tabebuia capitata*
mat^ho t^hot^ho — *Arrabidaea* sp.
mat^ho t^hot^ho — *Callichlamys latifolia*
maxahara hanaki — *Justicia pectoralis* var. *stenophylla*
maxahara sihi — *Stryphnodendron pulcherrimum*
maxopo mahi — *Amphirrhox surinamensis*
mayëpë xiki — *Cyperus articulatus*
mirama asihi — *Zizyphus cinnamomum*
misikirima hanaki — *Geonoma?* sp.
mōe hi — *Siparuna guianensis*
moka uku — *Marasmius cubensis*
mokamo si — *Bactris monticola*
mokuruma si — *Ischnosiphon arouma*
mokuruma si — *Ischnosiphon obliquus*
mominari usihi — *Pourouma ovata*
momo hi — *Micrandra rossiana*
momo t^hot^ho — *Cissus erosa*
mōra mahi — *Dacryodes peruviana*
morimoripë — *Begonia humilis*

morokoma t^hot^ho — *Philodendron* cf. *divaricatum*
mosipoima hanaki — *Piper francovilleanum*
moxima hi — *Inga alba*
mraka nahi — *Licania kunthiana*
naïra hi — *Chrysophyllum argenteum*
naharä hi — *Zanthoxylum pentandrum*
nahiri hanaki — *Zanthoxylum rhoifolium*
nai hi — *Manilkara bidentata*
nara uhi — *Copaifera* sp.
nara xihi — *Bixa orellana*
naxuruma ahi — *Quiina florida* vel aff.
naxuruma aki — *Costus* spp.
naxuruma t^hot^ho — *Passiflora coccinea*
në wāri hanaki — *Peperomia magnoliifolia*
nini kiki — *Renealmia floribunda*
ōema ahi — *Pourouma bicolor* ssp. *digitata*
ōha moki — *Cardiospermum* sp.
ohote hanaki — *Justicia pectoralis*
ohote kiki — *Cyperus articulatus*
oko hi — *Phyllanthus brasiliensis*
oko xiki — *Cyperus articulatus*
ōkorasi si — *Maximiliana maripa*
okoraxi hi — *Rinorea lindeniana*
Omoari si — *Renealmia floribunda*
operema axi hi — *Couma macrocarpa*
opo sihi — *Zollernia paraensis*
oru kiki wite — *Peperomia rotundifolia*
oruhe hi — *Rheedia benthamiana*
oruxi hi — *Anacardium giganteum*
paa hanaki — *Geonoma baculifera*
paara hi — *Anadenanthera peregrina*
paari maheko siki — *Piper bartlingianum*
paari mahekoki — *Piper demeraranum*
paari makasi hi — *Lacunaria jenmani*
paetea sihi — *Dipteryx odorata*
pahi hi — *Inga acreana*
paira ahi — *Swartzia* sp.
parapara hi — *Phyllanthus brasiliensis*
paraparamo nat^hoki — *Tanaecium nocturnum*
paro koxiki — *Dialium guianense*
paxo hi — *Martiodendron* sp.
paxo hwātemo hi — *Pouteria caimito*
paya hi — *Sclerolobium* sp.
pee nahe — *Nicotiana tabacum*
peesisima hi — *Perebea guianensis*
peima hi — *Chrysochlamys eberbaueri*
pihi wayorema amoki — *Filoboletus gracilis*
pirima ahi t^hot^ho — *Spondias mombin*
pirima hiki — *Andropogon bicornis*
pisa asi — *Calathea* cf. *majestica*
pitima hi — *Bellucia grossularioides*
poataa hi — *Inga pilosula*
pokara amoki — *Pleurotus flabellatus*
pokara mamo kasiki — *Solanum oocarpum*
poo hēt^ho hi — *Aspidosperma nitidum*

pooko hi	*Inga sarmentosa*
pora axi	*Posadaea sphaerocarpa*
pore hi	*Eugenia flavescens* vel aff.
pori pori tʰotʰo	*Clusia* spp.
poroa unahi	*Theobroma cacao*
poroma hanaki	*Xanthosoma sagittifolium*
poroma tʰotʰo	*Syngonium vellozianum*
poxe mamokasi hi	*Pouteria* sp.
prika aki	*Capsicum frutescens*
prōō masihi	*Theobroma microcarpum*
purunama usi	*Olyra latifolia*
puu si	*Saccharum officinarum*
puu tʰotʰo	*Philodendron solimoesensis*
puu tʰotʰo moki	*Tanaecium nocturnum*
rïa moxiririma hi	*Inga acuminata*
rāāsirima tʰotʰo	*Machaerium quinata*
rahaka mahi	*Aspidosperma* sp.
rai a	*Dioscorea* sp.
raina tihi	*Anaxagorea acuminata*
rapa hi	*Martiodendron* sp.
rapa mahi	*Nectandra* sp.
rasa si	*Bactris gasipaes*
raxana tihi	*Sterculia pruriens*
rëxë hi	*Perebea angustifolia*
rihuwari si	*Jacaratia digitata*
rioko si	*Mauritia flexuosa*
ripo tʰotʰo	*Philodendron hylaeae*
rokoari si	*Carica papaya*
romi hanaki	*Justicia pectoralis*
ruapa hi	*Caryocar villosum*
ruru asi	*Phenakospermum guyannense*
ruru hi	*Rhodostemonodaphne grandis*
saruma asi	*Calathea* sp.
seisei unahi	*Guatteria* spp.
si waima a	*Hippeastrum puniceum*
siiriama sihi	*Vismia guianensis*
sikãri a	*Iryanthera juruensis*
sikãri a	*Iryanthera laevis*
sina hi	*Sagotia racemosa*
tüwakarama tʰotʰo	*Bauhinia guianensis*
taitai ahi	*Apeiba membranacea*
tapra	*Caladium bicolor*
teosi hanaki	*Psammisia guianensis*
teria hi	*Maquira calophylla*
tʰomi koxi	*Orthoclada laxa*
tʰomira asihi	*Exellodendron barbatum*
tʰooro a	*Solandra grandiflora*
tʰoru hwãtemo hi	*Tabernaemontana sananho*
tʰua hanaki	*Justicia pectoralis*
tʰua mamo hi	*Aniba riparia*
tʰua mamoki	*Cyperus articulatus*
tihi hi	*Mouriri nervosa*
tihi kiki	*Caladium bicolor*
tihitihi nahi	*Pogonophora schomburgkiana*
tixo nahe	*Besleria laxiflora*
tokori hanaki	*Cecropia* aff. *peltata*
tokosi hanaki	*Phlebodium decumanum*
totori mamo hi	*Annona ambotay*
totori mamo hi	*Myrcia* sp.
totori mamoki	*Caladium bicolor*
toxa hi	*Inga paraensis*
uki sihi	*Zollernia paraensis*
uxirima amoki	*Favolus spathulatus*
wāha aki	*Dioscorea trifida*
wāha urihitʰeri aki	*Dioscorea trifida*
waihi hanaki	*Cymbopogon citratus*
waiporo unahi	*Theobroma subincanum*
waitʰiri kiki	*Cyperus articulatus*
wakamoxi kiki	*Cyperus articulatus*
wakōwakō axi tʰotʰo	*Mucuna urens*
wapo kohi	*Clathrotropis macrocarpa*
waraka ahi	*Pourouma melinonii*
warama si	*Geonoma deversa*
warapa kohi	*Protium* spp.
warapa kohi	*Protium spruceanum*
warë unasi	*Apeiba?* sp.
wari mahi	*Ceiba pentandra*
warihinama usihi	*Pourouma minor*
wāro uhi	*Couepia caryophylloides*
waxia hi	*Casearia javitensis*
wayawaya tʰotʰo	*Gouania frangulaefolia*
werehe tʰotʰo	*Abuta rufescens*
weri nahi	*Posoqueria latifolia*
werihisi hi	*Pradosia surinamensis*
weyeri hi	*Protium fimbriatum*
witari mahi	*Vismia cayennensis*
xīkī a	*Cecropia* spp.
xãã a	*Monstera adansonii*
xama mamoki	*Cyperus articulatus*
xama siosiki	*Miconia lateriflora*
xaraka a	*Gynerium sagittatum*
xaraka ahi	*Manilkara huberi*
xenoma a	*Dieffenbachia bolivarana*
xihini hi	*Licania* cf. *polita*
xiho xihi	*Cordia nodosa*
xikirima hanaki	*Pharus virescens*
xina āthe	*Lonchocarpus* cf. *chrysophyllus*
xinarū uhi	*Gossypium barbadense*
xirixiri ātʰe	*Banisteriopsis lucida*
xitokoma hi	*Chimarrhis* sp.
xitokoma hi	*Swartzia schomburgkii*
xitopari hi	*Jacaranda copaia*
xoko hwaka hi	*Tetragastris altissima*
xokopë amoki	*Favolus brasiliensis*
xokopë āthe	*Serjania grandifolia*
xoomo si	*Astrocaryum gynacanthum*
xopa hi	*Helicostylis tomentosa*
xōrahe tʰotʰo	*Plukenetia abutaefolia*

xōwa	*Caladium bicolor*	*yãpi uhi*	*Isertia hypoleuca*
xoxomo hi	*Caryocar pallidum,*	*yãre hi*	*Ecclinusa guianensis*
	C. glabrum	*yaremaxi hi*	*Brosimum utile*
xuhuturi unahi	*Herrania lemniscata*	*yãri tʰotʰo*	*Thoracocarpus bissectus*
xuu tʰotʰo	*Aristolochia disticha* vel aff.	*yarimonahe*	*Dioscorea piperifolia*
yɨpi hi	*Maprounea guianensis*	*yãwa xihi*	*Pouteria hispida*
yɨpi hi	*Sorocea muriculata* ssp.	*yaware nahasiki*	*Erechtites hieraciifolia*
	uaupensis	*yawë kiki*	*Cyperus articulatus*
yãkoana hi	*Virola elongata*	*yoasi hi*	*Vismia angusta*
yãma asiki	*Ananas* sp.	*yonomo si*	*Zea mays*
yao nahi	*Xylopia* sp.	*yopo unaki*	*Asplundia* sp.
yãpi mamo hi	*Casearia guianensis*	*yoroko axihi*	*Vataireopsis* cf.
yãpi mamohi	*Margaritaria nobilis*		*surinamensis*

* These species are similar in appearance, and may have been confused.

Appendix III

Species of fungi collected by Prance *et al.* among the Yanomami*

Species	How eaten	Habitat	Voucher	Source (location)
Collybia pseudocalopus (Henn.) Sing.	R	F	23612	Toototobi
Collybia subpruinosa (Murr.) Dennis	R	F, P	23662	Toototobi
Coriolus zonatus (Nees) Quélet	B	MP	21398	Auaris
Favolus brasiliensis (Fr.) Fr.	B	F	23605	Toototobi
	B,R	MP, F	21317	Auaris
			10526	Serra das Surucucus
Favolus brunneolus Berk. & Curt.	—	—	21318	Auaris
Favolus tesselatus Mont.	B	MP	20082	Auaris
	—	—	13615	Serra das Surucucus
Gymnopilus earlei Murr.	B, R	F	23607	Toototobi
Gymnopilus hispidellus Murr.	—	—	21550	Auaris
Hydnopolyporus palmatus (Hook. in Kunth) O. Fid.	B	MP	20083	Auaris
Lactocollybia aequatorialis Sing.	B	MP	21414	Auaris
Lentinus crinitus (L. ex Fr.) Fr.	B	MP	20024	Auaris
	B	MP	20015	Auaris
Lentinus glabratus Mont. in Sagra	B, C	MP	20084	Auaris
Lentinus velutinus Fr.	B	MP	21392	Auaris
Leucocoprinus cheimonoceps (Berk. & Curt.) Sing.	R	F	23663	Toototobi
Neoclitocybe bissiseda (Bres.) Sing.	—	—	10516	Serra das Surucucus
Panus rudis Fr.	B	MP	20016	Auaris
Pholiota bicolor (Speg.) Sing.	B	SF	21322	Auaris
Pleurotus concavus (Berk.) Sing.	B	—	20088	Auaris
Polyporus aquosus Henn.	C	F	21316	Auaris
Polyporus stipitarius Berk. & Curt.	—	—	10515	Serra das Surucucus
Polyporus tricholoma Mont.	B	MP	21313	Auaris

B = Boiled; R = Roasted; C = Crude (raw); F = Field; MP = Manihot (cassava) plantation; P = Plantation; SF = Secondary forest

* Taken from the tables in Prance (1984). Only those fungi which were identified to species have been listed here

INDEX